The Best American Science and Nature Writing 2007

GUEST EDITORS OF
THE BEST AMERICAN SCIENCE
AND NATURE WRITING

2000 DAVID QUAMMEN
2001 EDWARD O. WILSON
2002 NATALIE ANGIER
2003 RICHARD DAWKINS
2004 STEVEN PINKER
2005 JONATHAN WEINER
2006 BRIAN GREENE
2007 RICHARD PRESTON

The Best American Science and Nature Writing™ 2007

Edited and with an Introduction
by Richard Preston

Tim Folger, Series Editor

HOUGHTON MIFFLIN COMPANY
BOSTON · NEW YORK 2007

www.houghtonmifflinbooks.com

ISSN 1530-1508
ISBN-13: 978-0-618-72224-2 ISBN-10: 0-618-72224-6
ISBN-13: 978-0-618-72231-0 (pbk.) ISBN-10: 0-618-72231-9 (pbk.)

Printed in the United States of America

MP 10 9 8 7 6 5 4 3 2 1

"How to Get a Nuclear Bomb" by William Langewiesche. Appeared in the *Atlantic Monthly,* December 2006, and was revised by the author for his book *The Atomic Bazaar: The Rise of the Nuclear Poor* by William Langewiesche. Copyright © 2007 by William Langewiesche. Reprinted by permission of Farrar, Straus, and Giroux, LLC.

"The Effeminate Sheep" by Jonah Lehrer. First published in *Seed,* June/July 2006. Copyright © 2006 by Jonah Lehrer. Reprinted by permission of the author.

"Let There Be Light" by Michael D. Lemonick. First published in *Time,* September 4, 2006. Copyright © 2006 by Time Inc. Reprinted by permission.

"The Nature of Violence" by Jeffrey A. Lockwood. First published in *Orion,* January/February 2006. Copyright © 2006 by Jeffrey A. Lockwood. Reprinted by permission of the author.

"The Germs of Life" by Lynn Margulis and Emily Case. First published in *Orion,* November/December 2006. Copyright © 2006 by Lynn Margulis and Emily Case. Reprinted by permission of the authors.

"Neanderthal Man" by Steve Olson. First published in *Smithsonian,* October 2006. Copyright © 2006 by Steve Olson. Reprinted by permission of the author.

"Health Secrets from the Morgue" by Michael Perry. First published in *Men's Health,* April 2006. Copyright © 2006 by Michael Perry. Reprinted by permission of International Creative Management Inc.

"Hitler's Willing Archaeologists" by Heather Pringle. From *The Master Plan: Himmler's Scholars and the Holocaust* by Heather Pringle. Copyright © 2006 Heather Pringle. Reprinted by permission of Hyperion. All rights reserved. Headnote reprinted with permission of *Archaeology* magazine, 59:2 (copyright © the Archaeological Institute of America 2006).

"Sex, Lies, and Video Games" by Jonathan Rauch. First published in the *Atlantic Monthly,* November 2006. Copyright © 2006 by Jonathan Rauch. Reprinted by permission of the author.

"The Flu Hunter" by Michael Rosenwald. First published in *Smithsonian,* January 2006. Copyright © 2006 by Michael Rosenwald. Reprinted by permission of the author.

"Notes on the Space We Take" by Bonnie J. Rough. First published in *Ninth Letter,* Fall/Winter 2006–2007. Copyright © 2006 by Bonnie J. Rough. Reprinted by permission of the author.

"The Olfactory Lives of Primates" by Robert M. Sapolsky. First published in the *Virginia Quarterly Review,* Spring 2006. Copyright © 2006 by Robert M. Sapolsky. Reprinted by permission of the author.

"Ruffled Feathers" by John Seabrook. First published in *The New Yorker,* May 29, 2006. Copyright © 2006 by John Seabrook. Reprinted by permission of the author.

"In the Company of Bears" by Bill Sherwonit. First published in the *Anchorage Press,* July 13–July 19, 2006. Copyright © 2006 by Bill Sherwonit. Reprinted by permission of the author.

"The Rape of Appalachia" by Michael Shnayerson. First published in *Vanity Fair,* May 2006. Copyright © 2006 by Michael Shnayerson. Reprinted by permission of the author.

"First Soldier of the Gene Wars" by Meredith F. Small. First published in *Archaeology,* May/June 2006. Copyright © 2006 by Meredith F. Small. Reprinted by permission of the author.

"A Plan to Keep Carbon in Check" by Robert H. Socolow and Stephen W. Pacala.

Contents

Foreword

IF NOT FOR QUANTUM MECHANICS, an old friend of mine might never have tracked me down. I last saw Peter, who hailed from Liverpool, in the summer of 1974 in the middle of the North Sea oil fields. We were aboard the *Cherokee,* a 350-foot-long pipe-laying barge. I labored as a deckhand, Peter as an electrician. After a year of working twelve-hour shifts, seven days a week, I was eager to leave the barge and start college. Peter planned to travel around the world with his girlfriend, Pauline. We both assumed that not much time would pass before we'd meet again. So much for assumptions.

Fast-forward thirty-three years. (Did anyone fast-forward anything thirty-three years ago?) A few weeks ago, while routinely checking my morning e-mail, I found something completely unexpected in my in box: a message from Peter, our first contact since our days on the barge. He had used a search engine to find me — a quotidian tool that now seemed nearly miraculous to me in its power to erase decades. Without the Internet and personal computers, my friendship with Peter might have remained no more than a memory. And personal computers — not to mention the Internet — would not exist without the advances in electronics made possible by quantum mechanics.

When quantum mechanics was being developed in the 1920s and 1930s, the word *computers* referred to men and women employed to perform tedious calculations. Werner Heisenberg, Niels Bohr, Wolfgang Pauli, and the other illuminati who created the theory couldn't have imagined that their brainchild would lead to Google, YouTube, iTunes, or downloadable ringtones. None of them thought about the possible practical applications of their work. They were trying to understand the most basic properties of the universe. Yet today, by some estimates, roughly 30 percent of the gross national product of the United States derives from technologies based in one way or another on quantum mechanics.

If this anthology had been in publication seventy years ago, it might have contained an article about the strange new theory of quantum mechanics. But I doubt any such story would have anticipated how a seemingly arcane branch of science would transform the world, or how quickly most of us would come to take such revolutionary changes for granted, even as our due.

One of the joys of reading a collection like this is the opportunity to share the perspective of writers who take little for granted, who remind us that there is nothing ordinary about our world and that there is perhaps no better means for uncovering the unexpected than science. Our daily lives contain dangers and delights that most of us overlook. Take a ubiquitous product of modern life: plastics. Every bit of plastic ever made, Susan Casey tells us in her extraordinary story, still exists. The stuff is indestructible. Countless clumps of plastic waste pollute the world's oceans. Casey's narrative describes how plastic water bottles, polystyrene coffee cups, and other household detritus have entered the marine food chain, with staggering implications for our health.

If you happen to be reading this beneath an electric light, I refer you to Michael Shnayerson's disconcerting news about the hidden costs of our electricity. Most of the electricity generated in the United States comes from coal-fired power plants. And though coal mining has always been a dirty, dangerous business, I think you'll be shocked by Shnayerson's passionate account of what current mining practices are doing to the forests and people of Appalachia.

Along with the dangers that make some of these stories so gripping, there are, as I've mentioned, delights as well. Brian Doyle's "Fishering" is a poetic description of a fleeting wilderness encounter. Patricia Gadsby, in "Cooking for Eggheads," shows that a dash of science can add much to the appreciation of a meal, even for the most refined palates. Video game fans — and nonfans, for that matter — will enjoy Jonathan Rauch's "Sex, Lies, and Video Games." While reading Rauch's story, I remembered Pong, one of the world's first video games and, like its sophisticated descendants, another unexpected offspring of quantum theory. By today's hyperrealistic standards, Pong was extremely crude, a rather dull and slow video version of table tennis. But when it first appeared in the early 1970s, crowds lined up outside bars where the machines were installed. I know, because I waited in line to play, alongside my old friend Peter, who, I'm glad to report, did travel around the world with his girlfriend, Pauline. They are still together; happily, some things don't change.

So if any readers happen upon this collection some decades from now, what will strike them the most? Will they be astonished at what's not included? That something big and, to them, obvious is missing? Will they wonder how we failed to notice the potential of a technology or an obscure theory that changed the world? I'll bet they will.

Speaking of the future, I hope readers, writers, and editors will nominate their favorite articles for next year's anthology at http://timfolger.net/forums. The criteria for submissions, deadlines, and the address to which entries should be sent can be found in the "news and announcements" forum on my Web site. I also encourage readers to use the forums to leave feedback about the collection and to discuss all things scientific. The best way for publications to guarantee that their articles are considered for inclusion in the anthology is to place me on their subscription list, using the address posted in the forums.

It has been a pleasure to work with a gifted writer like Richard Preston. He has assembled a wonderfully diverse collection of stories. It's a pity guest-editorships don't last longer than one volume. Don't miss his latest book, *The Wild Trees: A Story of Passion and Daring.*

As always, I'm indebted to Amanda Cook and Will Vincent at Houghton Mifflin for their thoroughness and their uncanny ability to lasso people like Richard Preston. I'm also grateful to the late Niels Bohr, Werner Heisenberg, and the other physicists who created quantum theory, for their contribution in reuniting two old friends. I'm even more grateful to my beauteous wife, Anne Nolan, who is ever so much more enticing than Pong.

TIM FOLGER

Introduction

WHEN ASKED BY his students for a definition of good writing, John McPhee, the author of nonfiction books on subjects as diverse as nuclear bombs, citrus fruit, shad fishing, and continental drift, answers, "Good writing is where you find it." I took McPhee's journalism course at Princeton when I was graduate student in English literature there. At the time, I had a notion I would be an English professor, but I was becoming attracted to narrative nonfiction writing — the doing of it as distinct from the study of it. McPhee also said to his students, "A writer is someone who lacks the character not to write." Subsequently, I became a writer.

Good writing is where you find it. Often it seems to come out of a private obsession that the writer has with his or her material, which has gone beguilingly out of control. When you're moving through the swamp of creation, trying to slog your way to the end point in the drafting of a book or an article, you can come to feel that your obsession with the material is nothing less than a disgrace, and that nobody in their right mind would want to read what you're writing. In making the selections for this year's *Best American Science and Nature Writing*, I confess, I've been attracted to pieces in which the author displays a hint of obsession, especially if it involves a topic that's fresh, little known, or offbeat.

Having decided after graduate school that I would start doing nonfiction journalism, I ended up attempting to get published in *The New Yorker.* The magazine was then being edited by the physically small but editorially imposing William Shawn. He had a certain way of rejecting articles. He would send you a little slip of paper with a note on it, typed by himself with a mechanical typewriter, in which he said he thought you had written a "fine" piece, but, unfortunately, he couldn't use it. As everyone at the magazine knew, in William Shawn's lexicon, "fine" meant "bad." The first big thing I ever submitted to *The New Yorker* was a proposal to write a long article about astronomy. After a delay of about a year, I got a little typed note back from Mr. Shawn thanking me for my "very fine" proposal. (At least not many hopefuls got the "very fine" appellation from him.) He couldn't accept it, however, because, as he explained, he was no longer editor of *The New Yorker.* He had been forcibly retired by the magazine's owner, S. I. Newhouse. By that time, though, I was unable to stop writing about astronomy: my self-restraint had broken down.

The authors of the pieces selected for this volume share an intellectual passion for what they're writing about, a fascination with the subject matter as well with the human characters in their pieces, which is infectious to us as readers, and carries us into unknown worlds. They also have an ability to pull off some significant magic with words. It's been a delight, as an editor, to encounter really good, classy writing in places where you just sort of find it. I like geologist-journalist Bill Sherwonit's piece, "In the Company of Bears," published originally in the *Anchorage Press.* Sherwonit opens the piece on an islet near Shuyak Island, in the Kodiak Archipelago, in the Gulf of Alaska, where a 700-pound female brown bear, with cubs, charges and starts mauling ("engulfing," in Sherwonit's word) a guide named Sam while Sherwonit runs for his life. A little later we meet Sam standing there, unhurt, saying, "Thank goodness it was a friendly bear." It's an excellent opening for an excellent piece of nature writing about *Ursus arctos middendorffi,* or the Kodiak brown bear, one of the largest predators in nature — bigger than a grizzly. Sherwonit's piece is also a work of spiritual autobiography; it's about the author's relationship with the bears. The best science and nature writing, while seeing nature with precision, ultimately circles through human emotions and explores what we are as much as what nature is.

Michael Perry, in "Health Secrets from the Morgue," describes witnessing a medical autopsy of a man who died in his forties of natural but obscure causes. It's the best account of a medical autopsy I've ever read, as well as being a quasi-medieval reminder of what "lifestyle choices" can to for one's body. (" 'The liver should be reddish brown,' [the pathologist says of a tan-spotted liver], while running his finger over the surface. 'See how it's greasy? Almost oily? These are signs of alcohol damage.' ") Perry's piece gives us reasons that medical autopsies are important, and indeed, why it's a good idea to go to a hospital where autopsies are commonly done.

The term "science writing" has always seemed a little dubious to me. What's the definition of "science writing"? Is Henry David Thoreau's *Walden; or, Life in the Woods* an example of it? Is Herman Melville's *Moby Dick* (particularly the so-called cetacea chapters in that book, which describe the anatomy of the sperm whale and the methods of nineteenth-century whaling) another example of science writing? I think gloriously so, in both cases.

Science writing, at its best, imaginatively explores the intersection between the external world of nature and the internal world of the human spirit. Writing is a cultural act, an endeavor of the imagination. Narrative and story are ancient ways that humans have for passing knowledge from one generation to the next. Science writing is about the uneasy and often tortured relationship of the human species with nature. Nature — the dark energy, inflationary spacetime, birth and death of stars, particle interactions, the origin and evolution of life, the workings of the genetic code over immense reaches of time: all these things are the Other of nature; they seem to have an external reality that exists apart from our perception of them. If the human species had never come along, these things and processes, we feel, would still exist. If, in some perhaps nearly unnoticeable cosmic accident, the Earth were to vanish, it would not be missed: the Milky Way galaxy would still be evolving as an immense physical system containing a hundred billion stars and who knows what sort of dark matter.

And yet we do exist, and we have a central nervous system. With it we perceive the nature around us and inside us. We try to make sense of the universe, see patterns in it, and use it as well, toolmakers that we are. But we cannot ever be entirely a part of nature, for we stand apart as perceivers and analyzers. So often science is about breaking the shackles of belief and culture and achieving clear sight of the Other in nature, the system that makes no intuitive sense to the human mind, except that it is verifiable through an experiment or observation that can be repeated with the same result each time.

A writer who tackles science has to deal with this problem at an especially thorny level. A writer is an artist, or, to put it more modestly, a wordsmith who deals with culture. A scientist has a set of tools with which to characterize nature at a systematic level. A scientist uses the language of mathematics and the precision of technical language (the language found in a publication in *Nature* or *Science,* for example) to construct a formalized and communicable model of nature. A writer comes equipped with a different and more archaic set of tools. The writer must absorb Homer and Aristotle — that is, must grasp the methods of storytelling, the art of narrative, and the techniques of explanation, using language — to get across to general readers the immense mystery of the Other in nature.

The art of storytelling is old, embedded in thousands of years of culture. We don't know how old narrative really is. We don't really know when human vocalization developed enough complexity to engage in symbolic representation and metaphor. In the culture of the West, the Homeric poems stand among the earliest surviving, and still the best, examples of storytelling. What we can conjecture about the Homeric poems is that the poet (probably not someone named Homer) would speak a poem from memory and from a lifetime of practice at improvisation. It's a winter night in the Peloponnesus around 800 B.C.E. We see Homer standing in a great room, the megaron of a house, before a circular hearth, where a log fire is burning to coals. As she speaks the *Odyssey,* in a chant divided into a sixfold measure formed by the number and length of syllables, she plucks a small harp, a lyre. Or perhaps someone else is plucking a lyre to accompany her voice. The room is quiet, filled with intent faces in the firelight, faces of people from all walks of life — slaves, children, peasants, nobles — all listening. Some of listeners are highly educated people, though possibly not one of them knows how to read. Homer's story goes on for several nights in succession, and it's all about the world, the whole known world. The writer who's dealing with science must hold the attention of an educated audience, many of whom are illiterate in the language of science, and has to tell them about the whole world in a way they can't forget.

Strong narrative is useful in science and nature writing, where a story can provide propulsive force to the subject matter — to follow the life of a scientist and describe the scientist's obsessions helps to show readers why the scientific material may be important for all of us. Writing is linear, proceeding from one word to the next; one of the tricks in science and nature writing, which works well in certain kinds of pieces but is a hard trick to get right, is the delicate process of managing the line of prose as it moves from exposition to narrative back to exposition again. Technically, when you break a narrative to explain something, it's called a "set piece." The name tells what it does. It sits there, providing the reader with explanation of something. If you put too much narrative in a piece of writing about science, without enough exposition, the reader won't see the reason for the narrative. But if you start a narrative and then hang too much exposition on it, the exposition ends up as a load of wet laundry hanging on the line, and it drags the narrative down to the ground.

The pieces that stirred and lifted me often did so through character. At some basic level, all writing is about human character. This is also true of the expository essay, in which there is always a strong central character: the author of the essay, of course. For example, Neil deGrasse Tyson's "Delusions of Space Enthusiasts" sparkles with humor and paradox. The essay seems casual enough, but it's a sharp and witty look at Americans' delirium over space travel, and it resonates with Tyson's voice and personality.

In John Seabrook's "Ruffled Feathers" we meet another sort of character, Colonel Richard Meinertzhagen, the late, distinguished British ornithologist who amassed a huge collection of stuffed bird skins, which are in the British Natural History Museum and which he claimed to have collected himself. Turns out he stole a lot of them from museums or bought them on the sly and then intentionally mislabeled and faked them; he may also have murdered his wife, Anne, with a bullet to the brain, after she allegedly found out about his frauds. (After he inherited his wife's estate, Meinertzhagen took up with his seventeen-year-old cousin, Theresa Clay, who eventually became his scientific collaborator, helping him study parasites on birds.) Meinertzhagen's fraud was suspected or known by insiders in the clubby world of museum ornithology, who said nothing about it. Three laudatory biographies of Meinertzhagen were published, the most recent in 1998, none of which labeled him a faker and a thief. The other character in Seabrook's piece is the ornithologist Pamela Rasmussen, who recently traced Meinertzhagen's fraud and proved it while she was writing a monumental book on bird taxonomy. You can't help but relish reading about a scoundrel of a bird expert, as well as about the bird expert who revealed his crimes.

Heather Pringle's "Hitler's Willing Archaeologists" profiles the SS's top archaeologist, Dr. Assien Bohmers, a tall blond Dutch scholar who, working for SS chief Heinrich Himmler (who liked archaeology), took over important Cro-Magnon sites in Europe and attempted to show that they proved both the existence and the superiority of a "Nordic race" of Cro-Magnons. At the same time, SS archaeologists were involved with medical experiments on and murders of Jewish prisoners in concentration camps, as well as the looting of archaeological museums in order to engorge Himmler's personal collection with rare artifacts. One doesn't usually associate archaeologists with crimes against humanity, but there it is.

This year's selected pieces also provide a haul of more attractive characters. Among them, in Jonathan Rauch's "Sex, Lies, and Video Games," we get a portrait of two video game designers, Michael Mateas and Andrew Stern, who have been writing video games as interactive dramas, which are meant to unfold like a novel with a million possible endings. Naturally, the video-game industry is skeptical. "People love to blow shit up," as one executive told Jonathan Rauch. The games are nevertheless headed toward unbelievably powerful and real-seeming interactive drama, according to Rauch. Video games "will be as emotionally deep and meaningful to you as your dreams," he says, quoting a guru of gaming. Rauch's characters may never succeed, but his portrayal of them does.

Another thing I relish is clear exposition of an important idea, especially if it's counterintuitive, challenging, controversial, or hasn't been presented in such a way before. Here we don't need narrative or character; what we need is a good argument. "A Plan to Keep Carbon in Check," in *Scientific American,* by Robert H. Socolow and Stephen W. Pacala, who are both working scientists, has a quiet but strong voice and a good raison d'être. The authors ask the straightforward question, How, practically speaking, can the world reduce carbon emissions? If we know the climate is warming up because of human-caused carbon emissions, then if we begin acting right now, what can we do to reduce carbon emissions, and how effective will it be? In a straightforward, persuasive set of arguments, Socolow and Pacala show that carbon emissions can be lowered, and it can have a major effect.

Science journalist John Horgan's claim that scientists have already solved most of the major mysteries of nature has elicited a number of snorting dismissals from scientists who argue that we still know very little about nature. I tend to agree with them, but I have included his piece "The Final Frontier" here because he's a great arguer and provocateur. Horgan challenges the reader to think about what science really is and how it really proceeds. Plus, he seems to have ticked off a number of Nobel laureates with his ideas, and that can't be all bad.

William Langewiesche's "How to Get a Nuclear Bomb" is simply brilliant — flawless writing combined with "bomb-grade" reportage. He asks a simple question: If a terrorist group wanted to build a nuclear bomb, how would they actually do it, and where would they get their materials? Langewiesche leads us from a clan leader's drawing room in Turkish Kurdistan to remote towns in the Urals on the trail of nuclear bomb ingredients. Good stuff.

How is one supposed to write about science? Careful note-taking (which can involve the use of recording media other than a notebook, such as a voice recorder, a still camera, or a video camera) is an essential part of the process of good nonfiction writing about any subject. Note-taking should be followed up by fact-checking. Many scientists I've encountered while I've been researching a piece have been initially suspicious. Somewhere along the line they feel they got burned by a journalist. They believe they were misquoted or that their work was distorted or sensationalized in the media. Often they're offended when a journalist both oversimplifies their work and seems to pump up its significance. This can have the effect of making the scientist look like he or she is overclaiming — ascribing too much significance to a finding or to the scientist's own contribution to it — which a major sin in the scientific community. While there is no shortage of ego in the scientific world, the truth is that most working scientists probably feel that they, as people, are far less interesting than their result — what's cool is the thing they have found out about nature. What's important is nature. "I just don't see why you need to know how I eat Oreos," one scientist said to me when I wanted to know if he was a splitter or a chunker. (It was important because he seemed to eat a lot of Oreos, and that's a character detail that adds sparkle to the portrait; plus I didn't want to get my facts wrong.)

Scientists typically work in teams, and for many good reasons they want their collaborators to be acknowledged. Teamwork and no flashy personalities — that doesn't always fit well with the narrative needs of journalists, who are attracted like hoverflies to eccentric loners doing brilliant work that goes against the prevailing opinion. Scientists often qualify their findings with ambiguities or unknowns. Their findings may be only indicative, subject to more collection of data, and the implications may not be clear. "This work isn't finished, and there's a lot we *don't* know" — one hears that kind of statement all the time. I find that I often miss these qualifications during early interviews with a scientist — I'm too busy just trying to get the basics right.

Hence the need for fact-checking. All of the authors in this anthology clearly got involved with fact-checking, often, no doubt, through repeated, careful interviews; and many magazines these days fact-check the articles that appear in them anyway, as a policy. The result can be a cleaner, sharper piece of writing, with a texture of verisimilitude that feels solid to a reader. I do most of my fact-checking on the telephone. Fact-checking can end up involving the scientist-subject in the art of writing, which is fun (for me, anyway). When I'm dealing with scientists, I read passages and sentences aloud to them over the phone, sometimes again and again. The scientist reacts, while I make changes to the text in real time, scribbling on the draft with a pencil or typing on the computer. This can be time-consuming and frustrating, but in practice it improves the writing. Fact-checking also helps to establish a sense of trust with the scientist. It can convince the scientist that the journalist is as obsessed with the material as the scientist is.

At the beginning of my career, my "very fine" article about astronomy ended up a becoming a book, and while researching it,

I conducted a number of interviews with Maarten Schmidt, the Dutch-American astronomer who discovered the distance of quasars, brilliant cores of galaxies that contain black holes. Schmidt is a softspoken and agreeable man. I interviewed him at Palomar Observatory, in California, where he proved to be polite but not particularly communicative; he seemed edgy. During the night (while he was on an observing run to photograph galaxies using the 200-inch Hale Telescope), I noticed that he would disappear for periods of time. I thought he might be having digestive problems, so I didn't say anything. Finally I asked his colleagues, "Where did Maarten go?" "He's up on the catwalk," someone said. They explained that Maarten Schmidt liked to go up on the high catwalk of the dome of the Hale Telescope and stare at the stars, by himself. So I followed him to the catwalk to interview him, carrying my notebook and a flashlight (so that I could see what I was writing). I found Schmidt walking counterclockwise around the catwalk in near-total darkness. He didn't want me to turn on my light, even for a split second, because it would wipe out his night vision — his ability to see sixth-magnitude stars like grains of sand. I put my notes away and asked him what was on his mind.

In a quiet voice, he described a lifetime of mysteries and questions about the universe, questions he'd pursued for his entire scientific career, everything he'd pondered about the universe since childhood. He spoke of the worry and difficulty of his current research, even his nervousness that clouds would roll in and spoil the view, so he would lose a night of precious observation. He paced the catwalk in a state of thoughtful emotion. I was a twenty-something would-be writer trailing along behind him in the darkness, unable to take notes, able only to listen. Afterward, I rushed downstairs to a lighted room and scrawled everything I could remember in a notebook.

Months later, I drafted a passage describing Maarten Schmidt walking counterclockwise around the catwalk of the Hale Telescope and staring at the stars, and I tried to describe his thoughts as he had narrated them to me. That was where the fact-checking came in. I called him on the telephone several times and checked the developing passage word for word, more than once. As he spoke, I listened to the nuances of his Dutch-accented voice and made many small but important changes in the passage, until essentially what emerged from the fact-checking was a nonfiction interior monologue, a checkable version of a character's stream of thoughts as a novelist might portray them. Schmidt agreed, after this lengthy and probably very boring (for him) series of sessions, that the finished passage was "pretty much what goes through my mind when I'm looking at the stars." In other words, he had confirmed a repeatable result. In theory, another journalist could come along and question Maarten Schmidt about his thoughts and get more or less the same answer. But it wouldn't result in exactly the same collection of sentences, since writing is story, and storytelling is art, not science.

Possibly my favorite among all these selections is Jeffrey A. Lockwood's essay, "The Nature of Violence." It's a tour de force with a surprise ending that abruptly reveals the art of the piece. There is a moving portrayal of a scientist here. Lockwood, an entomologist and professor at the University of Wyoming, reflects on his observations and experiments with gryllacridids, "insects that look like a cross between a cricket and a grasshopper." Gryllacridids are uncompromisingly fierce. They will attack anything that seems threatening, no matter now large it is. They are programmed to attack; it's a selected trait. Lockwood describes one gryllacridid in his lab in Australia that he accidentally injured; the little creature bent over and ate the fat and entrails spilling out of its own body. And then Lockwood does something cool: he relates this alien-seeming organism to the human soul and draws a link between the Other and us.

What I find compelling in these stories about science is the odd interplay, the curious disjunction, between the awkward, humble, passionate, sometimes comical or even vile elements of the human self and the play of nature, the serene Other. In the end, science is not about facts and discoveries, it's about mystery. Science is about not knowing and wanting badly to know. Science is about flawed and complicated human beings trying to use whatever tools they've got, along with their minds, to see something strange and new. In that sense, writing about science is just another way of writing about the human condition.

RICHARD PRESTON

The Best American Science and Nature Writing 2007

PAUL BENNETT

In Rome's Basement

FROM *National Geographic*

LUCA PUSHES his head into the sewer, inhales, and grins. "It doesn't smell so bad in the cloaca today," he says, dropping himself feet first into a dark hole in the middle of the Forum of Nerva. Despite his optimism, the blackness emits a sickening aroma: a mélange of urine, diesel, mud, and rotting rat carcasses. In short, it smells just as you'd expect a 2,500-year-old continuously used sewer to smell. Below in the dark, tuff-vaulted cavern itself, things aren't much better. As Luca wades through water the color of army fatigues, stepping over fragments of temples and discarded travertine washed down over the ages, a diorama of modern life floats past: cigarette butts, plastic bags, plastic lighters, a baby pacifier, and a disturbingly large amount of stringy gray stuff that looks like toilet paper, although raw sewage isn't supposed to be flowing through here. At one turn, Luca points out a broken amphora, perhaps 2,000 years old, lying in the mud next to a broken Peroni beer bottle, perhaps a week old. Together they provide a striking testament to how long people have been throwing their garbage into the gutter of this city.

Luca Antognoli, forty-nine, works for Roma Sotterranea, a group of urban speleologists commissioned by the city to explore Rome's subterranean spaces — an amazing array of temples, roads, houses, and aqueducts buried by history since the fall of the Roman Empire. According to tradition, the Cloaca Maxima ("great drain"), which runs beneath the Roman Forum, was built in the sixth century B.C., making it one of the city's oldest — if not the oldest — surviving structures. So it is surprising to learn, as Luca winds his way through the sludge-filled passage under Via Cavour, that the cloaca has never been fully explored and mapped.

In real life Luca Antognoli is a surgeon, and he has warned us to be careful not to expose our skin to the water, a potent mix of street runoff and raw sewage. Earnest and wide-eyed, he has taken the danger seriously, covering every inch of his body with gloves, boots, hooded wind suit, and mask — all hermetically sealed with duct tape. He motions sharply at a conduit disgorging a surge of ocher liquid into the cavern that aerosolizes into a mist, sending members of the group into a frenzy of fitting masks over their faces.

He points out other conduits, some dumping clean water into the sewer from underground springs, some releasing dirty water. At one point we pass through a sloping section down which brown sludge purls. Beyond this dangerous obstacle lies a deep hole where, sometime during the past 2,000 years, the floor has washed out, forcing everyone to inch along an unseen precipice in chest-high, scum-covered water. A joker in the group observes that it looks like *schiuma,* the cocoa-like foam on Italian espresso.

At a pile of rubble — bones, pottery shards, and caked mud that nearly fill the entire space of the cloaca — the adventure comes to a halt. The sewer's barrel vault clearly reaches into the darkness beyond — one wonders how far.

Roma Sotterranea plans to send a remote-controlled robot to probe beyond the barrier; Luca expects to confirm that the great drain reaches the Baths of Diocletian, nearly a mile northeast. Who knows what treasures lie along the way, he says, noting that archaeologists had recently pulled a colossal head of Emperor Constantine from a sewer just like this, prompting speculation that the first Christian emperor may have been the victim of *damnatio memoriae,* as the practice of obliterating the memory of despised emperors was known in ancient Rome.

For Luca Antognoli, subterranean spaces like the Cloaca Maxima offer clues about how this city grew to rule an empire from the edge of Scotland to Baghdad, leaving its imprint indelibly on Western history.

A rivulet coming from the darkness flows down the rubble. Someone asks if it's dirty or clean. "It's very dirty," Luca says, eyeing the opening beyond, "but very important."

*

The cloaca, originally an open drain, was intentionally buried during the time of the Roman Republic, but most of what underlies Rome is there accidentally, buried by two millennia of sedimentation and urban growth.

"Rome has been rising for three thousand years," says Darius Arya, an archaeologist and director of the American Institute for Roman Culture. Much of Rome is situated in a floodplain, including the modern city center, known in antiquity as Campus Martius, at a bend of the Tiber River. Although the Romans put up levees, the city still flooded periodically, so they built upward, laying new structures and streets on earlier ones. "It was cost-effective, and it worked," Arya says. "We see the Romans jacking their city up two meters at a time, raising themselves above the water but also burying their past."

Today the city sits on layers of history 45 feet deep in places. But ironically, while the beguiling truth of Rome is that you can dig a hole anywhere within the 12-mile ring of walls that once enclosed the ancient city and find something of interest, comparatively little of this buried city has been excavated.

"I don't imagine more than ten percent has been documented," Robert Coates-Stevens says. An archaeology fellow with the British School at Rome, Coates-Stevens has been trying for a decade to piece together the topography of ancient Rome. During the 1800s, the Roman Forum was dug out — work that continues — but most ancient structures are still trapped under the traffic-clogged streets and office buildings of the contemporary city. "It's a heady feeling," Coates-Stevens says, "to think that all this still lies beneath our feet awaiting discovery."

In the 1920s and '30s, seized with this kind of excitement, Benito Mussolini razed sections of Rome's historic center, where medieval and Renaissance houses stood, to reveal the layers below — specifically anything dating back to the time of Emperor Augustus. (Mussolini liked to compare himself to Augustus and equated fascism with Pax Romana, the time of peace ushered in by Augustus.) By the 1980s this big-hole approach to archaeology had fallen out of favor, in part because of the financial challenge of protecting the ruins Mussolini had exposed from acid rain, smog, and vandalism. But curiosity about Rome is eternal, and so the vanguard of archaeology has shifted: archaeologists, and the speleologists they employ, are exploring ancient spaces from below, leaving the surface undisturbed.

Cristiano Ranieri pulls a dry suit over his head and fixes a full-lace respirator in place. Above, the hum of tourist traffic bounces off the travertine and brick surfaces of the Colosseum. But down here, among the maze of passages where gladiators would have waited and lifts would have raised lions, bears, and other exotic animals up to the action, the sound is muffled. Ranieri becomes visibly excited as he describes scuba diving beneath the Colosseum, explaining that this space, inside the 40-foot-deep "doughnut" foundation that holds up the rest of the structure, isn't even the bottom. He removes a steel plate from the floor to reveal a still body of dark gray water several yards below: the underbelly of the underbelly of the Colosseum.

Ranieri is scouting a new access point into the drain system for a future measuring project. In particular, he wants to know whether he will be able to carry a full underwater lighting rig into this hole or be forced to use portable flashlights. A well-lit swim below the Colosseum could change history.

Until three years ago, only a quarter of the conduits — the driest and most easily accessible — below the Colosseum had been explored. These simple drains, designed to whisk away storm water, date from the late first century A.D., when the Flavian emperors were building the Colosseum. Some ancient writers claimed the building was deliberately flooded for mock naval battles. But there was no evidence of the large waterworks needed to bring in the water.

Then, in October 2003, Ranieri, an archaeologist and speleologist with the superintendency of archaeology, made a startling discovery. Below the simple drains (and predating the Colosseum) were large conduits constructed by Emperor Nero to charge an artificial lake in his gardens. The conduits had obviously been reused by the architects of the Colosseum, most likely to pipe quantities of water in and out. For the first few years of its history, at least, the Colosseum, like many other theaters, was capable of being flooded.

Far more common than planned expeditions to reveal Rome's hidden secrets are the chance discoveries. A work crew digs a hole in the street and cracks into a hollow underground space. Speleologists are called in, and yet another astounding find alters the picture. Such was the case on the Oppian Hill two winters ago, when, after a period of intense rain, a hole spontaneously opened near a tree, exposing a matrix of underground rooms.

Marco Placidi, a coolheaded speleologist and a founder of Roma Sotterranea, was called in. Using ropes and harnesses, Placidi lowered himself into a dark, 40-foot-high room, which archaeologists believe was built sometime after Nero's nearby Domus Aurea ("golden house"), dating from A.D. 65, and sometime before the Baths of Trajan (circa A.D. 109), located above both of these structures.

This room, it turned out, is one of the best-preserved features from the Roman world, with meticulously flat brickwork and large arches. But for Placidi, the heart-thumping moment came halfway down, when, hanging in midair, he aimed his headlamp at the wall: on it was a mosaic, in perfect condition, showing a group of naked men harvesting and stamping grapes. The "Vendemmia" ("Grape Harvest"), as it was called, is some 10 feet long and made of minuscule, vividly colored bits of marble and other stone. "When I dropped down into this hole, I never imagined I'd see something like this," Placidi says. "It was an immense joy."

To an outsider, the randomness of such discoveries is shocking. But for Romans, it is quotidian. In the course of going about his business, someone somewhere bumps up against an artifact that hasn't seen the light of day for hundreds — or thousands — of years. Every year the city authorizes 13,000 requests for building permits, each of which requires archaeological evaluation. Construction of roads and sewers in Rome's ever-expanding suburbs is years behind because the overwhelming number of finds stops work and throws budgets into disarray.

The city of Rome has been trying hard to extend a sewer to the Appian area in the southeast for the past three years but has made little progress, according to Davide Mancini and Sergio Fontana. They run a cooperative of archaeology graduate students called Parsifal, contracted by the city's cash-strapped archaeology office to monitor the work of dozens of construction crews. At any time Parsifal may be overseeing up to twenty worksites, looking out for artifacts, reporting back to the government, and, if necessary, halting work to analyze finds. Traditional archaeology is painstakingly slow, but Parsifal's experts must be able to spot a precious object and assess its value in the seconds before the shovel plunges in.

On a typical day at the site, as Mancini and Fontana watch, a backhoe has to stop four times in a single half-hour. A harried graduate student jumps into the muddy pit and tosses up treasure: a lamp, several plates and bowls, small terra-cotta sculptures, and countless fragments of amphorae — much of which might date from the third or fourth century B.C.

"This zone is a mess," Davide says. He explains that work began here in June 2003, expecting the project would take a few months. But completion is still nowhere in sight.

A few minutes later the backhoe stops again. The student hops into the hole and sends up several very large pieces of amphorae, one of which has a glob of something stuck to it. Brightening, Sergio sniffs it. He says that it might be resin used to seal the amphora, a rare find. Davide disagrees. He thinks it might be incense, maybe from the amphora's reuse in medieval times. Regardless, the fragment goes into a plastic bag, which goes into a box next to dozens of other boxes that wait for the truck to take them to a warehouse. Meanwhile, the backhoe driver finishes his cigarette and asks permission to continue digging.

Interruptions such as these cannot be planned for in advance, in part because it is impossible to know what the underground has in store. (Ground-penetrating radar, which works well in rural settings, has difficulty differentiating complex, debris-filled soils in continuously inhabited places.) Beyond that, the length of any delay depends very much on the value assigned to what is found. Some things, like amphorae shards, can be quickly dismissed. Others, like buildings, may need to be sketched, measured, and otherwise documented. And occasionally, if something is unique, the Italian state may mandate that it be made accessible to the public.

This troubles many Romans. "No one wants the Beni Culturali knocking on their door," Robert Coates-Stevens says, referring to the state ministry that oversees archaeology. Traditionally, private property owners have been loath to report that errant column in the basement. But this could change as people come to think of having a piece of ancient history as an asset rather than a liability.

Five years ago Alda Fendi, scion of the Fendi fashion empire, bought a section of a Renaissance palace in central Rome just yards from the Column of Trajan. Her interior designer gutted the space for an art gallery, but in the course of the work, laborers digging in the basement discovered architectural footings of the Basilica Ulpia, a law court built by Emperor Trajan in A.D. 112 and attached to his forum. After a brief excavation and documentation, the state archaeologists recommended that part of the paving be restored and left visible. Fendi understood the importance of the find and envisioned incorporating the basilica into her gallery. She got permission to finance continued digging through the foundation of the palace and out under the piazza in front. The work eventually revealed a large section of the Basilica Ulpia, including several columns along with well-preserved flooring made of green and yellow marble and purple Phrygian marble.

"The marbles were an emotional find," Fendi says. "They are beautiful, and I knew immediately that, if possible, I wanted to use them in our gallery."

Urban speleologists like to joke about their work, poking fun at each other as they slog through sewers scaring each other with tales of rats the size of dogs and other nonsense. But as Marco Placidi and Adriano Morabito slip into their dry suits and descend the spiral staircase called La Chiocciola ("snail shell") in central Rome, they get very quiet and very serious. To them, this is sacred space. At the bottom, the steps drop off into the Aqua Virgo, an underground aqueduct carved through solid rock by Emperor Augustus's right-hand man, Marcus Agrippa, in 19 B.C. It travels more than 10 miles from a spring in Salone, source of the most constant pure water supply Romans have ever enjoyed. The aqueduct is still in use, feeding the Trevi Fountain and Gian Lorenzo Bernini's Fountain of the Four Rivers in Piazza Navona.

Roma Sotterranea surveyed the Aqua Virgo for the city's water utility several years ago, fixing its position with lasers, compasses, and the occasional pop to the surface for a GPS reality check. This day the group has returned to double-check earlier work and to float in dry suits through the conduit.

The water, at cheek level, comes directly from the spring and is pretty cold. The walls of the Aqua Virgo, though cut from solid stone, are remarkably regular, recalling the rectilinear blocks of the Cloaca Maxima. But the two spaces couldn't be more different. One brings pure, life-giving water, while the other takes away putrid effluent. If the key to ancient Rome is water, then these two systems form the anchors of a critical continuum.

As the group presses on toward the Trevi Fountain, Nick De Pace, a teacher of architecture at the Rhode Island School of Design on a Fulbright fellowship studying Rome's underground structures, points out the curved lines where excavators scratched their way through the solid tuff two thousand years ago. He describes a work crew using hand tools to chisel, by torchlight, a smooth, sloping conduit that looks to modern eyes as if it had been made by machine. For him, Rome's aqueducts and sewers are emblematic of the can-do spirit of Roman civilization. "Nobody thinks the sewers are that important," he says, examining a stalactite where calcium-rich groundwater is seeping into the aqueduct. "But for me they explain how and why Rome existed."

The exploring party comes to a wall of modern concrete where the original conduit has been truncated and its water diverted into a modern pipe. The pipe takes the flow downhill to the Trevi Fountain and beyond. What might it lead to? A new branch veering off to feed some buried ancient building whose discovery would bring the infrastructure of Rome into sharper focus?

"Rome is the biggest open-air museum in the world," says Darius Arya of the American Institute for Roman Culture. "There's so much to explore. I find it funny that people talk about diving to the bottom of the sea or climbing faraway peaks. Here's Rome, where we still don't know what's underneath."

SUSAN CASEY

Plastic Ocean

FROM *Best Life*

FATE CAN TAKE strange forms, and so perhaps it does not seem unusual that Captain Charles Moore found his life's purpose in a nightmare. Unfortunately, he was awake at the time, and 800 miles north of Hawaii in the Pacific Ocean.

It happened on August 3, 1997, a lovely day, at least in the beginning: Sunny. Little wind. Water the color of sapphires. Moore and the crew of *Alguita,* his 50-foot aluminum-hulled catamaran, sliced through the sea.

Returning to Southern California from Hawaii after a sailing race, Moore had altered *Alguita*'s course, veering slightly north. He had the time and the curiosity to try a new route, one that would lead the vessel through the eastern corner of a 10-million-square-mile oval known as the North Pacific subtropical gyre. This was an odd stretch of ocean, a place most boats purposely avoided. For one thing, it was becalmed. "The doldrums," sailors called it, and they steered clear. So did the ocean's top predators: the tuna, sharks, and other large fish that required livelier waters, flush with prey. The gyre was more like a desert — a slow, deep, clockwise-swirling vortex of air and water caused by a mountain of high-pressure air that lingered above it.

The area's reputation didn't deter Moore. He had grown up in Long Beach, 40 miles south of L.A., with the Pacific literally in his front yard, and he possessed an impressive aquatic résumé: deckhand, able seaman, sailor, scuba diver, surfer, and finally captain. Moore had spent countless hours in the ocean, fascinated by its vast trove of secrets and terrors. He'd seen a lot of things out there, things that were glorious and grand; things that were ferocious and humbling. But he had never seen anything nearly as chilling as what lay ahead of him in the gyre.

It began with a line of plastic bags ghosting the surface, followed by an ugly tangle of junk: nets and ropes and bottles, motor-oil jugs and cracked bath toys, a mangled tarp. Tires. A traffic cone. Moore could not believe his eyes. Out here in this desolate place, the water was a stew of plastic crap. It was as though someone had taken the pristine seascape of his youth and swapped it for a landfill.

How did all the plastic end up here? How did this trash tsunami begin? What did it mean? If the questions seemed overwhelming, Moore would soon learn that the answers were even more so, and that his discovery had dire implications for human — and planetary — health. As *Alguita* glided through the area that scientists now refer to as the "Eastern Garbage Patch," Moore realized that the trail of plastic went on for hundreds of miles. Depressed and stunned, he sailed for a week through bobbing, toxic debris trapped in a purgatory of circling currents. To his horror, he had stumbled across the twenty-first-century leviathan. It had no head, no tail. Just an endless body.

"Everybody's plastic, but I love plastic. I want to be plastic." This Andy Warhol quote is emblazoned on a 6-foot-long magenta and yellow banner that hangs — with extreme irony — in the solar-powered workshop in Moore's Long Beach home. The workshop is surrounded by a crazy Eden of trees, bushes, flowers, fruits, and vegetables, ranging from the prosaic (tomatoes) to the exotic (cherimoyas, guavas, chocolate persimmons, white figs the size of baseballs). This is the house in which Moore, fifty-nine, was raised, and it has a kind of open-air earthiness that reflects his '60s-activist roots, which included a stint in a Berkeley commune. Composting and organic gardening are serious business here — you can practically smell the humus — but there is also a kidney-shaped hot tub surrounded by palm trees. Two wet suits hang drying on a clothesline above it.

This afternoon Moore strides the grounds. "How about a nice fresh boysenberry?" he asks, and plucks one off a bush. He's a striking man wearing no-nonsense black trousers and a shirt with official-looking epaulettes. A thick brush of salt-and-pepper hair frames his intense blue eyes and serious face. But the first thing you notice about Moore is his voice, a deep, bemused drawl that becomes animated and sardonic when the subject turns to plastic pollution. This problem is Moore's calling, a passion he inherited from his father, an industrial chemist who studied waste management as a hobby. On family vacations, Moore recalls, part of the agenda would be to see what the locals threw out. "We could be in paradise, but we would go to the dump," he says with a shrug. "That's what we wanted to see."

Since his first encounter with the Garbage Patch nine years ago, Moore has been on a mission to learn exactly what's going on out there. Leaving behind a twenty-five-year career running a furniture-restoration business, he has created the Algalita Marine Research Foundation to spread the word of his findings. He has resumed his science studies, which he'd set aside when his attention swerved from pursuing a university degree to protesting the Vietnam War. His tireless effort has placed him on the front lines of this new, more abstract battle. After enlisting scientists such as Steven B. Weisberg, Ph.D. (executive director of the Southern California Coastal Water Research Project and an expert in marine environmental monitoring), to develop methods for analyzing the gyre's contents, Moore has sailed *Alguita* back to the Garbage Patch several times. On each trip the volume of plastic has grown alarmingly. The area in which it accumulates is now twice the size of Texas.

At the same time, all over the globe there are signs that plastic pollution is doing more than blighting the scenery; it is also making its way into the food chain. Some of the most obvious victims are the dead seabirds that have been washing ashore in startling numbers, their bodies packed with plastic: things like bottle caps, cigarette lighters, tampon applicators, and colored scraps that, to a foraging bird, resemble baitfish. (One animal dissected by Dutch researchers contained 1,603 pieces of plastic.) And the birds aren't alone. All sea creatures are threatened by floating plastic, from whales down to zooplankton. There's a basic moral horror in seeing the pictures: a sea turtle with a plastic band strangling its shell into an hourglass shape; a humpback towing plastic nets that cut into its flesh and make it impossible for the animal to hunt. More than a million seabirds, 100,000 marine mammals, and countless fish die in the North Pacific each year, either from mistakenly eating this junk or from being ensnared in it and drowning.

Bad enough. But Moore soon learned that the big, tentacled balls of trash were only the most visible signs of the problem; others were far less obvious, and far more evil. Dragging a fine-meshed net known as a manta trawl, he discovered minuscule pieces of plastic, some barely visible to the eye, swirling like fish food throughout the water. He and his researchers parsed, measured, and sorted their samples and arrived at the following conclusion: by weight, this swath of sea contains six times as much plastic as it does plankton.

This statistic is grim — for marine animals, of course, but even more so for humans. The more invisible and ubiquitous the pollution, the more likely it will end up inside us. And there's growing — and disturbing — proof that we're ingesting plastic toxins constantly, and that even slight doses of these substances can severely disrupt gene activity. "Every one of us has this huge body burden," Moore says. "You could take your serum to a lab now, and they'd find at least one hundred industrial chemicals that weren't around in 1950." The fact that these toxins don't cause violent and immediate reactions does not mean they're benign: scientists are just beginning to research the long-term ways in which the chemicals used to make plastic interact with our own biochemistry.

In simple terms, plastic is a mix of monomers linked together to become polymers, to which additional chemicals can be added for suppleness, inflammability, and other qualities. When it comes to these substances, even the syllables are scary. For instance, if you're thinking that perfluorooctanoic acid (PFOA) isn't something you want to sprinkle on your microwave popcorn, you're right. Recently the Science Advisory Board of the Environmental Protection Agency (EPA) upped its classification of PFOA to a likely carcinogen. Yet it's a common ingredient in packaging that needs to be oil- and heat-resistant. So while there may be no PFOA in the popcorn itself, if PFOA is used to treat the bag, enough of it can leach into the popcorn oil when your butter deluxe meets your superheated microwave oven that a single serving spikes the amount of the chemical in your blood.

Other nasty chemical additives are the flame retardants known as polybrominated diphenyl ethers (PBDEs). These chemicals have been shown to cause liver and thyroid toxicity, reproductive problems, and memory loss in preliminary animal studies. In vehicle interiors, PBDEs — used in moldings and floor coverings, among other things — combine with another group called phthalates to create that much-vaunted "new-car smell." Leave your new wheels in the hot sun for a few hours, and these substances can "off-gas" at an accelerated rate, releasing noxious byproducts.

It's not fair, however, to single out fast food and new cars. PBDEs, to take just one example, are used in many products, including computers, carpeting, and paint. As for phthalates, we deploy about a billion pounds of them a year worldwide despite the fact that California recently listed them as a chemical known to be toxic to our reproductive systems. Used to make plastic soft and pliable, phthalates leach easily from millions of products — packaged food, cosmetics, varnishes, the coatings of timed-release pharmaceuticals — into our blood, urine, saliva, seminal fluid, breast milk, and amniotic fluid. In food containers and some plastic bottles, phthalates are now found with another compound called bisphenol A (BPA), which scientists are discovering can wreak stunning havoc in the body. We produce 6 billion pounds of that each year, and it shows: BPA has been found in nearly every human who has been tested in the United States. We're eating these plasticizing additives, drinking them, breathing them, and absorbing them through our skin every single day.

Most alarming, these chemicals may disrupt the endocrine system — the delicately balanced set of hormones and glands that affects virtually every organ and cell — by mimicking the female hormone estrogen. In marine environments, excess estrogen has led to *Twilight Zone*–esque discoveries of male fish and seagulls that have sprouted female sex organs.

On land, things are equally gruesome. "Fertility rates have been declining for quite some time now, and exposure to synthetic estrogen — especially from the chemicals found in plastic products — can have an adverse effect," says Marc Goldstein, M.D., director of the Cornell Institute for Reproductive Medicine. Dr. Goldstein also notes that pregnant women are particularly vulnerable: "Prenatal exposure, even in very low doses, can cause irreversible damage in an unborn baby's reproductive organs." And after the baby is born, he or she is hardly out of the woods. Frederick vom Saal, Ph.D., a professor at the University of Missouri at Columbia who specifically studies estrogenic chemicals in plastics, warns parents to "steer clear of polycarbonate baby bottles. They're particularly dangerous for newborns, whose brains, immune systems, and gonads are still developing." Dr. vom Saal's research spurred him to throw out every polycarbonate plastic item in his house, and to stop buying plastic-wrapped food and canned goods (cans are plastic-lined) at the grocery store. "We now know that BPA causes prostate cancer in mice and rats, and abnormalities in the prostate's stem cell, which is the cell implicated in human prostate cancer," he says. "That's enough to scare the hell out of me." At Tufts University, Ana M. Soto, M.D., a professor of anatomy and cellular biology, has also found connections between these chemicals and breast cancer.

As if the potential for cancer and mutation weren't enough, Dr. vom Saal states in one of his studies that "prenatal exposure to very low doses of BPA increases the rate of postnatal growth in mice and rats." In other words, BPA made rodents fat. Their insulin output surged wildly and then crashed into a state of resistance — the virtual definition of diabetes. They produced bigger fat cells, and more of them. A recent scientific paper Dr. vom Saal coauthored contains this chilling sentence: "These findings suggest that developmental exposure to BPA is contributing to the obesity epidemic that has occurred during the last two decades in the developed world, associated with the dramatic increase in the amount of plastic being produced each year." Given this, it is perhaps not entirely coincidental that America's staggering rise in diabetes — a 735 percent increase since 1935 — follows the same arc.

This news is depressing enough to make a person reach for the bottle. Glass, at least, is easily recyclable. You can take one tequila bottle, melt it down, and make another tequila bottle. With plastic, recycling is more complicated. Unfortunately, that promising-looking triangle of arrows that appears on products doesn't always signify endless reuse; it merely identifies which type of plastic the item is made from. And of the seven different plastics in common use, only two of them — PET (labeled with #1 inside the triangle and used in soda bottles) and HDPE (labeled with #2 inside the triangle and used in milk jugs) — have much of an aftermarket. So no matter how virtuously you toss your chip bags and shampoo bottles into your blue bin, few of them will escape the landfill — only 3 to 5 percent of plastics are recycled in any way.

"There's no legal way to recycle a milk container into another milk container without adding a new virgin layer of plastic," Moore says, pointing out that because plastic melts at low temperatures, it retains pollutants and the tainted residue of its former contents. Turn up the heat to sear these off, and some plastics release deadly vapors. So the reclaimed stuff is mostly used to make entirely different products, things that don't go anywhere near our mouths, such as fleece jackets and carpeting. Therefore, unlike recycling glass, metal, or paper, recycling plastic doesn't always result in less use of virgin material. It also doesn't help that fresh-made plastic is far cheaper.

Moore routinely finds half-melted blobs of plastic in the ocean, as though the person doing the burning realized partway through the process that this was a bad idea and stopped (or passed out from the fumes). "That's a concern as plastic proliferates worldwide, and people run out of room for trash and start burning plastic — you're producing some of the most toxic gases known," he says. The color-coded bin system may work in Marin County, but it is somewhat less effective in subequatorial Africa or rural Peru.

"Except for the small amount that's been incinerated — and it's a very small amount — every bit of plastic ever made still exists," Moore says, describing how the material's molecular structure resists biodegradation. Instead, plastic crumbles into ever-tinier fragments as it's exposed to sunlight and the elements. And none of these untold gazillions of fragments is disappearing anytime soon: even when plastic is broken down to a single molecule, it remains too tough for biodegradation.

Truth is, no one knows how long it will take for plastic to biodegrade, or return to its carbon and hydrogen elements. We only invented the stuff 144 years ago, and science's best guess is that its natural disappearance will take several more centuries. Meanwhile, every year we churn out about 60 billion tons of it, much of which becomes disposable products meant only for a single use. Set aside the question of why we're creating ketchup bottles and six-pack rings that last for half a millennium and consider the implications of it: plastic never really goes away.

*

Ask a group of people to name an overwhelming global problem, and you'll hear about climate change, the Middle East, or AIDS. No one, it is guaranteed, will cite the sloppy transport of nurdles as a concern. And yet nurdles, lentil-sized pellets of plastic in its rawest form, are especially effective couriers of waste chemicals called persistent organic pollutants, or POPs, which include known carcinogens such as DDT and PCBs. The United States banned these poisons in the 1970s, but they remain stubbornly at large in the environment, where they latch on to plastic because of its molecular tendency to attract oils.

The word itself — *nurdles* — sounds cuddly and harmless, like a cartoon character or a pasta for kids, but what it refers to is most certainly not. Absorbing up to a million times the level of POP pollution in their surrounding waters, nurdles become supersaturated poison pills. They're light enough to blow around like dust, to spill out of shipping containers, and to wash into harbors, storm drains, and creeks. In the ocean, nurdles are easily mistaken for fish eggs by creatures that would very much like to have such a snack. And once inside the body of a bigeye tuna or a king salmon, these tenacious chemicals are headed directly to your dinner table.

One study estimated that nurdles now account for 10 percent of plastic ocean debris. And once they're scattered in the environment, they're diabolically hard to clean up (think wayward confetti). At places as remote as Rarotonga, in the Cook Islands, 2,100 miles northeast of New Zealand and a twelve-hour flight from L.A., they're commonly found mixed with beach sand. In 2004, Moore received a $500,000 grant from the state of California to investigate the myriad ways in which nurdles go astray during the plastic manufacturing process. On a visit to a polyvinyl chloride (PVC) pipe factory, as he walked through an area where railcars unloaded ground-up nurdles, he noticed that his pant cuffs were filled with a fine plastic dust. Turning a corner, he saw windblown drifts of nurdles piled against a fence. Talking about the experience, Moore's voice becomes strained and his words pour out in an urgent tumble: "It's not the big trash on the beach. It's the fact that the whole biosphere is becoming mixed with these plastic particles. What are they doing to us? We're breathing them, the fish are eating them, they're in our hair, they're in our skin."

Though marine dumping is part of the problem, escaped nurdles and other plastic litter migrate to the gyre largely from land. That polystyrene cup you saw floating in the creek, if it doesn't get picked up and specifically taken to a landfill, will eventually be washed out to sea. Once there, it will have plenty of places to go: the North Pacific gyre is only one of five such high-pressure zones in the oceans. There are similar areas in the South Pacific, the North and South Atlantic, and the Indian Ocean. Each of these gyres has its own version of the Garbage Patch, as plastic gathers in the currents. Together these areas cover 40 percent of the sea. "That corresponds to a quarter of the Earth's surface," Moore says. "So twenty-five percent of our planet is a toilet that never flushes."

It wasn't supposed to be this way. In 1865, a few years after Alexander Parkes unveiled a precursor to manmade plastic called Parkesine, a scientist named John W. Hyatt set out to make a synthetic replacement for ivory billiard balls. He had the best of intentions: save the elephants! After some tinkering, he created celluloid. From then on, each year brought a miraculous recipe: rayon in 1891, Teflon in 1938, polypropylene in 1954. Durable, cheap, versatile — plastic seemed like a revelation. And in many ways it was. Plastic has given us bulletproof vests, credit cards, slinky spandex pants. It has led to breakthroughs in medicine, aerospace engineering, and computer science. And who among us doesn't own a Frisbee?

Plastic has its benefits; no one would deny that. Few of us, however, are as enthusiastic as the American Plastics Council. One of its recent press releases, titled "Plastic Bags — A Family's Trusted Companion," reads: "Very few people remember what life was like before plastic bags became an icon of convenience and practicality — and now art. Remember the 'beautiful' [sic] swirling, floating bag in *American Beauty*?"

Alas, the same ethereal quality that allows bags to dance gracefully across the big screen also lands them in many less desirable places. Twenty-three countries, including Germany, South Africa, and Australia, have banned, taxed, or restricted the use of plastic bags because they clog sewers and lodge in the throats of livestock. Like pernicious Kleenex, these flimsy sacks end up snagged in trees and snarled in fences, becoming eyesores and worse: they also trap rainwater, creating perfect little breeding grounds for disease-carrying mosquitoes.

In the face of public outrage over pictures of dolphins choking on "a family's trusted companion," the American Plastics Council takes a defensive stance, sounding not unlike the NRA: plastics don't pollute, people do.

It has a point. Each of us tosses about 185 pounds of plastic per year. We could certainly reduce that. And yet — do our products have to be quite so lethal? Must a discarded flip-flop remain with us until the end of time? Aren't disposable razors and foam packing peanuts a poor consolation prize for the destruction of the world's oceans, not to mention our own bodies and the health of future generations? "If 'more is better' and that's the only mantra we have, we're doomed," Moore says, summing it up.

Oceanographer Curtis Ebbesmeyer, Ph.D., an expert on marine debris, agrees. "If you could fast-forward ten thousand years and do an archaeological dig . . . you'd find a little line of plastic," he told the *Seattle Times* last April. "What happened to those people? Well, they ate their own plastic and disrupted their genetic structure and weren't able to reproduce. They didn't last very long because they killed themselves."

Wrist-slittingly depressing, yes, but there are glimmers of hope on the horizon. Green architect and designer William McDonough has become an influential voice, not only in environmental circles but among Fortune 500 CEOs. McDonough proposes a standard known as "cradle to cradle" in which all manufactured things must be reusable, poison-free, and beneficial over the long haul. His outrage is obvious when he holds up a rubber ducky, a common child's bath toy. The duck is made of phthalate-laden PVC, which has been linked to cancer and reproductive harm. "What kind of people are we that we would design like this?" McDonough asks. In the United States, it's commonly accepted that children's teething rings, cosmetics, food wrappers, cars, and textiles will be made from toxic materials. Other countries — and many individual companies — seem to be reconsidering. Currently McDonough is working with the Chinese government to build seven cities using "the building materials of the future," including a fabric that is safe enough to eat and a new, nontoxic polystyrene.

Thanks to people like Moore and McDonough, and media hits such as Al Gore's *An Inconvenient Truth*, awareness of just how hard we've bitch-slapped the planet is skyrocketing. After all, unless we're planning to colonize Mars soon, this is where we live, and none of us would choose to live in a toxic wasteland or to spend our days getting pumped full of drugs to deal with our haywire endocrine systems and runaway cancer.

None of plastic's problems can be fixed overnight, but the more we learn, the more likely it is that eventually wisdom will trump convenience and cheap disposability. In the meantime, let the cleanup begin. The National Oceanographic & Atmospheric Administration (NOAA) is aggressively using satellites to identify and remove "ghost nets," abandoned plastic fishing gear that never stops killing. (A single net recently hauled up off the Florida coast contained more than 1,000 dead fish, sharks, and one loggerhead turtle.) New biodegradable starch- and corn-based plastics have arrived, and Wal-Mart has signed on as a customer. A consumer rebellion against dumb and excessive packaging is afoot. And in August 2006, Moore was invited to speak about "marine debris and hormone disruption" at a meeting in Sicily convened by the science adviser to the Vatican. This annual gathering, called the International Seminars on Planetary Emergencies, brings scientists together to discuss mankind's worst threats. Past topics have included nuclear holocaust and terrorism.

The gray plastic kayak floats next to Moore's catamaran, *Alguita,* which lives in a slip across from his house. It is not a lovely kayak; in fact, it looks pretty rough. But it's floating, a sturdy, 8-foot-long two-seater. Moore stands on *Alguita*'s deck, hands on hips, staring down at it. On the sailboat next to him, his neighbor, Cass Bastain, does the same. He has just informed Moore that he came across the abandoned craft yesterday, floating just offshore. The two men shake their heads in bewilderment.

"That's probably a six-hundred-dollar kayak," Moore says, adding, "I don't even shop anymore. Anything I need will just float by." (In his opinion, the movie *Cast Away* was a joke — Tom Hanks could've built a village with the crap that would've washed ashore during a storm.)

Watching the kayak bobbing disconsolately, it is hard not to wonder what will become of it. The world is full of cooler, sexier kayaks. It is also full of cheap plastic kayaks that come in more attractive colors than battleship gray. The ownerless kayak is a lummox of a boat, 50 pounds of nurdles extruded into an object that nobody wants but that'll be around for centuries longer than we will.

And as Moore stands on deck looking into the water, it is easy to imagine him doing the same thing 800 miles west, in the gyre. You can see his silhouette in the silvering light, caught between ocean and sky. You can see the mercurial surface of the most majestic body of water on Earth. And then below, you can see the half-submerged madhouse of forgotten and discarded things. As Moore looks over the side of the boat, you can see the seabirds sweeping overhead, dipping and skimming the water. One of the journeying birds, sleek as a fighter plane, carries a scrap of something yellow in its beak. The bird dives low and then boomerangs over the horizon. Gone.

RICHARD CONNIFF

For the Love of Lemurs

FROM *Smithsonian*

ON A STEEP SLOPE, hip deep in bamboo grass, in the heart of the Madagascar rain forest she saved, Patricia Wright is telling a story. "Mother Blue is probably the oldest animal in this forest," she begins. "She was the queen of group one, and she shared her queendom with what I think was her mother."

The animals she is describing are lemurs, primates like us. They are the unlikely product of one of nature's reckless little experiments: all of them — more than fifty living lemur species — derive from a few individuals washed from the African mainland into the Indian Ocean more than 60 million years ago. The castaways had the good luck to land on Madagascar, an island the size of Texas 250 miles off the southeast coast of Africa. And there they have evolved in wild profusion.

Wright, a late-blooming primatologist from the State University of New York at Stony Brook, has made lemurs her life, tracking bamboo lemurs and sifaka lemurs that live in a handful of social groups in Ranomafana National Park. The story she is telling, to a work party from the volunteer group Earthwatch, is one episode in a running saga from twenty years of field research in Madagascar. If her tone evokes a children's story, that may be apt. Wright is a matriarchal figure, with straight auburn hair framing a round face, slightly protuberant eyes under padded eyelids, and a quick, ragged grin. The business of conservation has made her adept at popularizing her lemurs, using all the familiar plotlines of wicked stepmothers, families broken up and reunited, love, sex, and murder.

A female sifaka lemur perches on a branch over Wright's head.

The graceful creature, a little bigger than a house cat, has a delicate, foxlike snout and plush black fur with a white patch on her back. Her long limbs end in skeletal fingers, curved for gripping branches, with soft, leathery pads at the tips. She turns her head, her stark, staring, reddish orange eyes glowing like hot coals. Then she bounds away in a series of leaps, a dancer in perfect partnership with the trees.

Wright first visited the town of Ranomafana in 1986, basically because she needed a bath. She was looking for the greater bamboo lemur, a species no one had seen in decades. Ranomafana had hot springs — and also a rain forest that was largely intact, a rarity on an island where the vast majority of forest had been destroyed. In the steep hills outside town, Wright spotted a bamboo lemur and started to track it, the first step in getting skittish wild animals to tolerate human observers. "You have to follow them and follow them and follow them, and they're very good at hiding," she says. "It's kind of fun to try to outwit an animal. When they decide that you're boring, that's when you've won."

The lemur Wright followed turned out to be an entirely new species, the golden bamboo lemur, which even locals said they had not seen before. (Wright shares credit for the discovery with a German researcher working in the area at the same time.) On a return trip, she also found the greater bamboo lemur she'd originally been looking for.

As Wright was beginning a long-term study in Ranomafana of both the bamboo lemurs and the sifakas in 1986, she came face-to-face with a timber baron with a concession from Madagascar's Department of Water and Forests to cut down the entire forest. Wright decided to try and preserve the lemurs' habitat. She was married, raising a young daughter, and employed at Duke University as a new faculty member. Friends warned that letting "this conservation stuff" distract her from research would hurt her career. "But I couldn't have it on my conscience," she says now, "that a species I had discovered went extinct because I was worried about getting my tenure."

Over the next few years, she pestered the timber baron so relentlessly that he abandoned the area. She lobbied government officials to designate Ranomafana as the nation's fourth national park, which they did in 1991, protecting 108,000 acres, an area five times the size of Manhattan. She also raised millions of dollars, much of it from the U.S. Agency for International Development, to fund the park. She oversaw the hiring of local villagers, construction of trails, and training of staff. She sent out teams to build schools and to treat diseases such as elephantiasis and roundworm, which were epidemic around the park. Her work won her a MacArthur Foundation "genius" grant, and Stony Brook wooed her away from Duke with a job offer that allowed her to spend even more time in Madagascar.

Along the way, Wright found time to get to know her lemurs as individuals, particularly the sifakas in five territorial social groups, each of which had three to nine lemurs. Pale Male, in group two, for instance, "was a great animal, very perky," she tells the volunteers. "He would play all the time with his sister, roughhouse around, go to the edges of the territory. And then one day, Pale Male disappeared. A lemur's lost call is a mournful whistle, and his sister gave it all day long." Pale Male had moved away to sifaka group three for an interlude of lemur bliss with the resident female, Sky Blue Yellow, producing a son named Purple Haze.

Lemurs typically sleep on the upper branches of trees. The fossa (pronounced "foosa"), a nocturnal mongoose, has a knack for finding them there. It creeps up a tree, its lean body pressed close to the bark, then leaps out and catches a lemur by the face or throat with its teeth. After a fossa struck one night, Sky Blue Yellow was gone. Pale Male, badly battered, soon also disappeared, leaving behind his two-year-old son, Purple Haze. Six months passed by the time Pale Male came back, bringing a new female into group three, and Wright was there to witness the reunion with Purple Haze. "That baby was so excited to see that father, and that father was so excited, and they just groomed and groomed and groomed."

Ranomafana, it turned out, was home to more than a dozen lemur species, all with behaviors worth studying. Wright went on to build an independent research station there called Centre ValBio (short for a French phrase meaning "valuing biodiversity"), which now employs more than eighty people and accommodates up to thirty students and researchers.

A few prominent academics say privately that Wright has not produced enough solid science, or trained enough students from Madagascar as full-time scientists, given the funding she has received. (Wright points to more than three hundred publications from research at Ranomafana.) Some conservationists complain that she steers initiatives to Ranomafana, sometimes at the expense of other parts of the island. "A lot of people are jealous of her," says Conservation International president Russ Mittermeier, who gave Wright the grant that brought her to Ranomafana. "But, boy, give me one hundred Pat Wrights and we could save a lot of primates."

Wright was a Brooklyn social worker when her career as a primatologist got its start with a purchase she describes now as "almost a sin." Before a Jimi Hendrix concert at the Fillmore East in Manhattan, Wright and her husband visited a nearby pet shop. A shipment had just arrived from South America, including a male owl monkey, says Wright, "and I guess I fell in love with that monkey."

Selling wild-caught monkeys is illegal today. But this was 1968, and the monkey, which was named Herbie, took up residence in the apartment where the Wrights also kept a large iguana, a tokay gecko, and a parrot. Monkey and parrot soon developed a mutual loathing. One night the monkey "made a leap for the parrot, and by the time we got the lights on he was poised with his mouth open about to bite the back of its neck." The parrot was sent to live with a friend.

Wright began to read everything she could about Herbie's genus, *Aotus,* nocturnal monkeys native to South and Central America. After a few years, she decided to find a mate for him. She took a leave of absence from her job and headed to South America for three months with her husband. Since no one wanted Herbie as a houseguest, he had to go too.

"I thought Herbie would be excited to see his own kind," Wright says of the females she eventually located in a village on the Amazon. But he regarded the female with an enthusiasm otherwise reserved for the parrot. Wright ended up chasing the two of them around a room to corral them into separate cages. Later this menagerie moved into a twenty-five-cent-a-day room in Bogotá. "I think the truth is, it was twenty-five cents an *hour* because it was a bordello. They thought it was hilarious to have this couple with two monkeys."

Back in New York, both Wright and the female owl monkey gave birth a few years later to daughters. Herbie turned into a doting father, returning his infant to its mother only for feeding. Wright stayed home with her own baby while her husband worked, and dreamed about someday discovering "what makes the world's only nocturnal monkey tick." Meanwhile, she sent off hapless letters — Brooklyn housewife yearns to become primatologist — to Dian Fossey, Jane Goodall, and the National Geographic Society.

Eventually she discovered that Warren Kinzey, an anthropologist at the City University of New York, had done fieldwork on another South American monkey species. Wright prevailed on Kinzey to talk with her about how to study monkeys, and she took careful notes: "Leitz 7×35 binoculars, Halliburton case, waterproof field notebook. . ." Then she persuaded a philanthropist from her hometown of Avon, New York, to pay for a research trip to study *Aotus* monkeys in South America.

"Don't go!" said Kinzey when Wright phoned to say goodbye. An article had just arrived on his desk from a veteran biologist who had been unable to follow *Aotus* at night even with the help of radio collars. "You don't have a radio collar," said Kinzey. "I don't think you should waste your money."

But Wright was undaunted. She'd been spending summers at a family cottage on Cape Cod, following her two monkeys as they wandered at night through the local forest. "It was just fun to see the things they would do in the middle of the night. They loved cicadas, and there was a gypsy moth outbreak one year and they got fat. They saw flying squirrels." So she told Kinzey, "I think I can do it without radio collars, and I've just bought a ticket, so I *have* to go."

A few days later she and her family climbed out of a bush plane in Puerto Bermudez, Peru, where her daughter Amanda, age three, shrieked at the sight of a Campa tribesman with face paint and a headdress. Wright said, "*¿Donde está el hotel turista?*" ("Where is the tourist hotel?"), and everybody within earshot laughed. The family moved in with some farmers before heading out into the field.

The local guides were nervous about going into the rain forest at night to help her hunt for owl monkeys. So Wright headed out alone, leaving behind a Hansel-and-Gretel trail of brightly colored flagging tape. She got lost anyway and began to panic at the thought of deadly fer-de-lance snakes and jaguars. "And then I hear this familiar sound, and it was an owl monkey. And I thought, okay, I can't act like I'm scared to death. I'll act like a primatologist. There are fruits dropping down in four places, so there are probably four monkeys. And I just started writing anything so I didn't have to think."

Near dawn she heard animals stampeding toward her, and she scrambled up a tree for safety. "I heard this sound above me, and it was an owl monkey scolding and urinating and defecating and saying, 'What are you doing in my territory?' And by the time he finished this little speech, it was daylight. And then he went into this tree and his wife followed right behind him, and I though, Oh, my god, that's their sleep tree."

She wrapped the tree with tape, "like a barber pole," so she could find it again, and made her way to camp. Six months later, back in the United States, she presented Kinzey with her study and got it published in a leading primatology journal. She also applied to graduate school in anthropology. In her second week of studies at the City University of New York, Wright and her husband separated.

The mother of all lemurs — the castaway species that somehow found its way to Madagascar — was probably a small, squirrel-like primate akin to the modern-day bush baby in central Africa. Prosimians (a name literally meaning pre-monkey, now used as a catch-all category for lemurs, lorises, and bush babies) tend to have proportionally smaller brains than their cousins, the monkeys and apes, and they generally rely more on scent than vision. There are now ring-tailed lemurs, red-bellied lemurs, golden-crowned lemurs, and black-and-white ruffed lemurs — so many different lemurs that Madagascar, with less than half a percent of the Earth's land surface, is home to about 15 percent of all primate species.

Among other oddities, the population includes lemurs that pollinate flowers, lemurs with incisors that grow continuously like a rodent's, lemurs that hibernate — unlike any other primate — and lemurs in which only the females seem to hibernate. The smallest living primates are mouse lemurs, able to fit in the palm of a human hand. An extinct lemur as big as a gorilla roamed the island until about 350 years ago. Lemurs species also display every possible social system, from polygyny (one male with multiple female partners) to polyandry (one female with multiple males) to monogamy.

Females are usually in charge. Males acknowledge the female's dominance with subtle acts of deference. They wait till she finishes eating before going into a fruit tree. They step aside when she approaches. They cede her the best spot in the roosting tree at night.

Female dominance remains one of the great unsolved mysteries of lemur behavior. Food sources are scattered on Madagascar, and highly seasonal. It may be that females need to control the limited supply to meet the nutritional demands of pregnancy and lactation. Big, tough, high-maintenance males would likely consume too many calories, Wright theorizes, and provide too little compensatory protection against a flash-in-the-night predator like the fossa. But whatever the explanation, the lemur system of low-key female leadership has become a source of deep, playful empathy for Wright.

Dominant females don't usually practice the sort of relentless aggression that occurs in male-dominated species such as baboons, macaques, and chimpanzees, she says. They typically commit only about one aggressive act every other day, and "they do it expeditiously. They run up and bite or cuff the individual, and it's very effective. They don't do a lot of strutting around saying, 'I'm the greatest.'" For every aggressive act, females engage in perhaps fifty bouts of friendly grooming, according to Wright's observations. In fact, grooming is so important to lemurs that it has shaped the evolution of their teeth. Whereas our lower canines and incisors stand upright, for biting and tearing, theirs stick straight out and have evolved into a fine-toothed comb plate, for raking through one another's hair.

Wright herself exerts dominance in the benign style of lemurs. "Zaka," she says one afternoon, taking aside one of her best fieldworkers for a sort of verbal grooming. "I have to tell you about how important you are. When we were looking at all the data from the survey you did, it was very nice, very nice." She is also a shrewd consensus builder, adept at winning local support. When she sends a student into the field, she urges him to hire local villagers as porters and guides, so they will see that the park can put money in their pockets. "I didn't know how to make a national park," Wright says. "What I did was brainstorm with the Malagasy [as people from

Madagascar are known] here and with the people in the Department of Water and Forests. It was always a group effort. They had to be a part of it, or it wasn't going to work at all."

Given her sense of identification with female leadership among lemurs, right was shocked when she learned recently that her greater bamboo lemurs have a dark secret. "Listen to them!" Wright cries out one morning on Trail W, where her lemurs are violently shredding the bark from towering bamboo stems. "They talk all the time. They crack open bamboo all the time. How in the world could I have had such a hard time following them for so many years?"

Female greater bamboo lemurs spend much of their day chewing through the hard outer surface of giant bamboo stems, till the pieces of stripped bark hang down like broken sticks of dry spaghetti. What the lemurs want is the edible pith, which looks about as appetizing as rolled vinyl. It also contains stinging hairs and, in young shoots, a small jolt of cyanide. Having adapted to digest that poison lets the species exploit bamboo, an otherwise underutilized resource.

"The female is using her teeth to open these bamboo culms, really working — and the male isn't there," says Wright. "And all of a sudden you hear this big squabbling noise, and the male appears just as she opens up the bamboo, and he displaces her and takes it from her!" The thought leaves her aghast. "*This is unheard of in Madagascar!* Then he moves on and takes away the bamboo from the next female."

At first Wright and graduate student Chia Tan thought they were simply seeing bad behavior by one beastly male. Then a new male came in and did the same thing, forcing the researchers to contemplate the possibility that the greater bamboo lemur may be the only male-dominated lemur species. Wright and Tan theorize that the females cannot hear anything over the racket of their own chewing; they need the male to patrol the perimeter and alert them to danger. But they pay the price at feeding time. "It's beautiful to watch," says Wright, "it's *horrible* to watch."

In another corner of the park, sifaka group three is feeding in a rahiaka tree, and Wright is talking about Mother Blue, the lemur for whom she has always felt the deepest empathy. During the first decade of Wright's work at Ranomafana, Mother Blue gave birth every other year, the normal pattern for sifakas. She raised two of her offspring to maturity, a good success rate for a lemur. Though female lemurs can live for more than thirty years, they produce relatively few offspring, most of which die young.

Mother Blue, says Wright, was not just a good mother but also a loving companion to her mate Old Red. "They groomed each other, they sat next to each other, they cared about each other." But Old Red eventually disappeared, and in July 1996, says Wright, a new female arrived in group one. Lemurs are by and large peaceful, but they still display the usual primate fixations on rank and reproductive opportunity. Male interlopers sometimes kill infants to bring their mothers back into mating condition. Female newcomers may also kill babies, to drive a rival mother out of a territory. Soon after the new female appeared, Mother Blue's newborn vanished. Then Mother Blue herself went into exile.

"I arrived a few months later and saw Mother Blue on the border between group one and group two, just sitting there looking depressed," says Wright. "I thought, this is what happens to old females. They get taken over by young females and just die."

Despite continuing deforestation elsewhere in Madagascar, satellite photographs indicate that Ranomafana remains intact. Partly because of the success there, Madagascar now has eighteen national parks. President Marc Ravalomanana has pledged to triple the amount of open space under government protection by 2008. Wright, among her other ambitions, hopes to establish a wildlife corridor stretching 90 miles south from Ranomafana. She also still yearns to find out what makes different species tick.

At the rahiaka tree, for instance, Earthwatch volunteers are keeping track of the lemurs as they feed on reddish fruit about the size of an acorn. The edible part, a rock-hard seed, is buried in a ball of gluey latex inside a tough, leathery husk. It doesn't seem to discourage the lemurs. One of them hangs languidly off a branch, pulling fruit after fruit into its mouth, which is rimmed white with latex. The sound of seeds being crunched is audible on the ground, where Wright watches with evident satisfaction.

It turns out Wright was mistaken about Mother Blue. The old female lemur did not simply go into exile and die. Instead, she has moved into group three and taken up with Pale Male's son, Purple

Haze, a decidedly younger male. The two of them have a three-year-old, also feeding in the tree, and a one-year-old, roaming nearby. Wright is delighted with the way things have worked out. (She has also taken up with another male: her second husband, Jukka Jernvall, a Finnish biologist.)

Mother Blue, whom Wright says is probably twenty-eight years old now, has worn teeth. The Earthwatchers are recording how much she eats and how many bites it takes her. They're also supposed to collect scat samples containing broken seed remnants, to see how well she digests it. Someone squeamishly points out where droppings have just fallen in the thick grass. Wright wades in. She grabs a couple of fresh pellets with her bare hands and bags them for analysis back in the lab. Then she turns and leads her group uphill, deeper into the Ranomafana forest. "There is nothing more exciting than finding a new thing nobody knows," says Wright. "You won't believe it, but everything hasn't already been discovered."

ALISON HAWTHORNE DEMING

The Rabbit on Mars

FROM *Isotope*

DO OTHER ANIMALS tell jokes? Perhaps the play of young animals in which they practice what will later become life-enhancing skills — the stalk, the pounce, the thrashing — is not so unlike what human beings do in attempting to learn what they will need to survive in the future. Of course, we live and direct our lives so much in the mind, so much in the richness and folly and, yes, beauty of what our minds can create, that our play often takes the form of jokes, a linguistic version of play. One of my recent favorites was the Rabbit on Mars. In 2004 the National Aeronautics and Space Administration landed two rovers, research-gathering devices, on our neighboring planet, the one with soil so bloody with iron that it was named after the Roman god of war. How much more benign is our view: let's do some geologic study of the soil and find out what it's really made of and whether any forms of life might once have inhabited the place, might (miraculous to say!) have blown loose in a cosmic wind and drifted here to our spinning globe, seeding everything we know, including the great and troubling argument between religion and science.

The two research devices are named Spirit and Opportunity, as if the project were intended, at last, to create a team from these often opposing forces. It has been an awe-inspiring experience to watch these little emissaries of our curiosity make their journey through space and land like bouncing balls in a place so far away we cannot imagine the distance — though we can cross it — then release themselves from their protective shells in response to messages beamed from Earth. Errors in the software? No problem — new instructions are beamed from home base, and the little brain is reconfigured, the rover rolling off its platform, drilling into Martian rock, sending snapshots back home through space. The scientists and engineers have worked in a collaboration as musical and passionate as a symphony orchestra to accomplish this, and their joy is beautiful to behold. How the knowledge will be used by Opportunity, if you will forgive my appropriating these names and returning them to their Earth-bound meanings, remains a cautionary tale. No green-thinking poet could celebrate transporting the culture of obsessive consumerism to another planet. It is how the knowledge will be used by Spirit that draws me to the curious phenomenon of the Rabbit on Mars.

Along with the panoramic images of barren rusty soil and rock circulating on the Internet — that collective unconscious of the technological that hovers over the surface of Earth — came a fuzzy image of an object that appears to have very long, erect, and pointed ears. It was graced with the caption "A Rabbit on Mars?" It is hardly the first time that human beings have projected their imaginations out into space. It isn't even the first time that the projections have taken the form of a rabbit. The Maya saw a leaping rabbit, the special pet of the moon goddess, where Americans see the man in the moon, and Japanese moon-gazers see a rabbit making a rice cake. Of course, everyone got the joke: there are no Martian bunnies hopping up to welcome our team of digital explorers. It turned out the rabbit image taken by Spirit was a tatter of the protective balloon that had helped cushion the spacecraft's landing. A momentary illusion, though it struck the imagination with silly and pleasurable force. Spirit, of course, will be struck when and if science finds life on other planets. And many astronomers and evolutionary biologists agree that it really is a matter of when, not if, now that we know planets are not an entirely exceptional phenomenon in the universe, and therefore the conditions suitable for life may not be limited to those on Earth. How far we have come from "War of the Worlds," as our scientific instruments have extended our eyes into far space. These eyes have altered our picture of what we will find there, so that we can now imagine the extraterrestrial not as mechanical warrior sent to conquer our planet but as a benign, fuzzy, harmless, and familiar creature, a vegetarian, an animal said to deliver baskets of candies to children at Eastertime.

Is the appeal of this image an instruction to ourselves? Can it calm us, to assuage the fears that our sense of our selves and our world may soon be sent reeling? What will it mean, even if all we find out there are microbes living in ice? Will our God be their God? Will Life itself become our God? Will we be humbled into a greater reverence and hunger for knowledge, as people were by the Copernican revolution? Will life become a dime-a-dozen happenstance to be owned, manipulated, destroyed, and devoured with an attitude of dismissal even greater than we've accomplished here on our gorgeous home planet? How could we know what rearrangement our spirits will undergo when such knowledge comes to us? What better guide for us — if only in play — than the animals who have been with us from the start, real and imagined, the animals who live in us as the matter of our genes and the spirit of our imaginations, who live with us as our teachers and companions and neighbors, the animals who were our first gods in the childhood of humanity.

BRIAN DOYLE

Fishering

FROM *Ecotone*

IN THE WOODS here in Oregon there is a creature that eats squirrels like candy, can kill a pursuing dog in less than a second, and is in the habit of deftly flipping over porcupines and scooping out the meat as if the prickle-pig were a huge and startled breakfast melon. This riveting creature is the fisher, a member of the mustelid family that includes weasels, otter, mink, badgers, ferrets, marten, and — at the biggest and most ferocious end of the family — wolverines. Sometimes called the pekan or fisher-cat, the fisher can be 3 feet long (with tail) and can weigh as much as 12 pounds. Despite its stunning speed and agility, it is best known not as an extraordinary athlete of the thick woods and snowfields, but as the bearer of a coat so dense and lustrous that it has been sought eagerly by trappers for thousands of years; this is one reason the fisher is so scarce pretty much everywhere it used to live.

Biologist friends of mine tell me there are only two "significant" populations of fisher in Oregon — one in the Siskiyou Mountains in the southwest, called the Klamath population, and the other in the Cascade Mountains south of Crater Lake, called the Cascade population. All of the rare sightings of fisher in Oregon in recent years have been in these two areas. In the northwest coastal woods where I occasionally wander, biologists tell me firmly, there are no fishers and there have been none for more than fifty years.

I am a guy who wanders around looking for nothing in particular, which is to say everything; in this frame of mind I have seen many things, in many venues urban and suburban and rural, and while ambling in the woods I have seen marten kits and three-legged elk and secret beds of watercress and the subtle dens of foxes. I have found thickets of wild grapevines, and secret jungles of salmonberries, and stands of huckleberries so remote and delicious that it is a moral dilemma as to whether or not I should leave a map behind for my children when the time comes for me to add to the compost of the world.

Suffice it to say that I have been much graced in these woods, but to see a fisher was not a gift I expected. Yet recently I found loose quills on the path and then the late owner of the quills, with his or her conqueror atop the carcass staring at me.

I do not know if the fisher had ever seen a human being before; it evinced none of the usual sensible caution of the wild creature confronted with *Homo violencia,* and it showed no inclination whatsoever to retreat from its prize. We stared at each other for a long moment and then I sat down, thinking that a reduction of my height and a gesture of repose might send the signal that I was not dangerous and had no particular interest in porcupine meat. Plus, I remembered that a fisher can slash a throat in less than a second.

Long minutes passed. The fisher fed, cautiously. I heard thrushes and wrens. There were no photographs or recordings, and when the fisher decided to evanesce, I did not take casts of its tracks or claim the former porcupine as evidence of fisherness. I just watched and listened and now I tell you. I don't have any heavy message to share. I was only a witness: where there are no fishers, there was a fisher. It was a stunning creature — alert, attentive, accomplished, unafraid. I think maybe there is much where we think there is nothing. Where there are no fishers, there was a fisher. Remember that.

HELEN FIELDS

Dinosaur Shocker!

FROM *Smithsonian*

SHE IS NOT your typical paleontologist. Neatly dressed in blue Capri pants and a sleeveless top, long hair flowing over her bare shoulders, Mary Schweitzer sits at a microscope in a dim lab, her face lit only by a glowing computer screen showing a network of thin, branching vessels. That's right, blood vessels. From a dinosaur. "Ho-ho-ho, I am excite-e-e-e-d," she chuckles. "I am, like, *really* excited."

After 68 million years in the ground, a *Tyrannosaurus rex* found in Montana was dug up, its leg bone was broken in pieces, and fragments were dissolved in acid in Schweitzer's laboratory at North Carolina State University in Raleigh. "Cool beans," she says, looking at the image on the screen.

It was big news indeed last year when Schweitzer announced she had discovered blood vessels and structures that looked like whole cells inside that *T. rex* bone — the first observation of its kind. The finding amazed colleagues, who had never imagined that even a trace of still-soft dinosaur tissue could survive. After all, as any textbook will tell you, when an animal dies, soft tissues such as blood vessels, muscle, and skin decay and disappear over time, while hard tissues like bone may gradually acquire minerals from the environment and become fossils. Schweitzer, one of the first scientists to use the tools of modern cell biology to study dinosaurs, has upended the conventional wisdom by showing that some rock-hard fossils tens of millions of years old may have remnants of soft tissues hidden away in their interiors. "The reason it hasn't been discovered before is no right-thinking paleontologist would do what

Mary did with her specimens. We don't go to all this effort to dig this stuff out of the ground to then destroy it in acid," says dinosaur paleontologist Thomas Holtz, Jr., of the University of Maryland. "It's great science." The observations could shed new light on how dinosaurs evolved and how their muscles and blood vessels worked. And the new findings might help settle a long-running debate about whether dinosaurs were warm-blooded or cold-blooded — or both.

Meanwhile, Schweitzer's research has been hijacked by "young earth" creationists, who insist that dinosaur soft tissue couldn't possibly survive millions of years. They claim her discoveries support their belief, based on their interpretation of Genesis, that the Earth is only a few thousand years old. Of course, it's not unusual for a paleontologist to differ with creationists. But when creationists misrepresent Schweitzer's data, she takes it personally: she describes herself as "a complete and total Christian." On a shelf in her office is a plaque bearing an Old Testament verse: "For I know the plans I have for you," declares the Lord, "plans to prosper you and not to harm you, plans to give you hope and a future."

It may be that Schweitzer's unorthodox approach to paleontology can be traced to her roundabout career path. Growing up in Helena, Montana, she went through a phase when, like many kids, she was fascinated by dinosaurs. In fact, at age five she announced she was going to be a paleontologist. But first she got a college degree in communicative disorders, married, had three children, and briefly taught remedial biology to high schoolers. In 1989, a dozen years after she graduated from college, she sat in on a class at Montana State University taught by paleontologist Jack Horner, of the Museum of the Rockies, now an affiliate of the Smithsonian Institution. The lectures reignited her passion for dinosaurs. Soon after, she talked her way into a volunteer position in Horner's lab and began to pursue a doctorate in paleontology.

She initially thought she would study how the microscopic structure of dinosaur bones differs depending on how much the animal weighs. But then came the incident with the red spots.

In 1991, Schweitzer was trying to study thin slices of bones from a 65-million-year-old *T. rex*. She was having a hard time getting the slices to stick to a glass slide, so she sought help from a molecular biologist at the university. The biologist, Gayle Callis, happened to take the slides to a veterinary conference, where she set up the ancient samples for others to look at. One of the vets went up to Callis and said, "Do you know you have red blood cells in that bone?" Sure enough, under a microscope, it appeared that the bone was filled with red disks. Later, Schweitzer recalls, "I looked at this and I looked at this and I thought, this can't be. Red blood cells don't preserve."

Schweitzer showed the slide to Horner. "When she first found the red blood cell–looking structures, I said, Yep, that's what they look like," her mentor recalls. He thought it was possible they were red blood cells, but he gave her some advice: "Now see if you can find some evidence to show that that's not what they are."

What she found instead was evidence of heme in the bones — additional support for the idea that they were red blood cells. Heme is a part of hemoglobin, the protein that carries oxygen in the blood and gives red blood cells their color. "It got me real curious as to exceptional preservation," she says. If particles of that one dinosaur were able to hang around for 65 million years, maybe the textbooks were wrong about fossilization.

Schweitzer tends to be self-deprecating, claiming to be hopeless at computers, lab work, and talking to strangers. But colleagues admire her, saying she's determined and hardworking and has mastered a number of complex laboratory techniques that are beyond the skills of most paleontologists. And asking unusual questions took a lot of nerve. "If you point her in a direction and say, Don't go that way, she's the kind of person who'll say, Why? — and she goes and tests it herself," says Gregory Erickson, a paleobiologist at Florida State University. Schweitzer takes risks, says Karen Chin, a University of Colorado paleontologist. "It could be a big payoff or it could just be kind of a ho-hum research project."

In 2000, Bob Harmon, a field crew chief from the Museum of the Rockies, was eating his lunch in a remote Montana canyon when he looked up and saw a bone sticking out of a rock wall. That bone turned out to be part of what may be the best-preserved *T. rex* in the world. Over the next three summers, workers chipped away at the dinosaur, gradually removing it from the cliff face. They called it B. rex in Harmon's honor and nicknamed it Bob. In 2001 they encased a section of the dinosaur and the surrounding dirt in plaster to protect it. The package weighed more than 2,000 pounds, which turned out to be just above their helicopter's capacity, so they split it in half. One of B. rex's leg bones was broken into two big pieces and several fragments — just what Schweitzer needed for her micro-scale explorations.

It turned out Bob had been misnamed. "It's a girl and she's pregnant," Schweitzer recalls telling her lab technician when she looked at the fragments. On the hollow inside surface of the femur, Schweitzer had found scraps of bone that gave a surprising amount of information about the dinosaur that made them. Bones may seem as steady as stone, but they're actually constantly in flux. Pregnant women use calcium from their bones to build the skeleton of a developing fetus. Before female birds start to lay eggs, they form a calcium-rich structure called medullary bone on the inside of their leg and other bones; they draw on it during the breeding season to make eggshells. Schweitzer had studied birds, so she knew about medullary bone, and that's what she figured she was seeing in that *T. rex* specimen.

Most paleontologists now agree that birds are the dinosaurs' closest living relatives. In fact, they say that birds *are* dinosaurs — colorful, incredibly diverse, cute little feathered dinosaurs. The theropod of the Jurassic forests lives on in the goldfinch visiting the backyard feeder, the toucans of the tropics, and the ostriches loping across the African savanna.

To understand her dinosaur bone, Schweitzer turned to two of the most primitive living birds: ostriches and emus. In the summer of 2004, she asked several ostrich breeders for female bones. A farmer called, months later. "Y'all still need that lady ostrich?" The dead bird had been in the farmer's backhoe bucket for several days in the North Carolina heat. Schweitzer and two colleagues collected a leg from the fragrant carcass and drove it back to Raleigh.

As far as anyone can tell, Schweitzer was right: Bob the dinosaur really did have a store of medullary bone when she died. A paper published in *Science* last June presents microscope pictures of medullary bone from ostrich and emu side by side with dinosaur bone, showing near-identical features.

In the course of testing a B. rex bone fragment further, Schweitzer asked her lab technician, Jennifer Wittmeyer, to put it in weak acid, which slowly dissolves bone, including fossilized bone — but not soft tissues. One Friday night in January 2004, Wittmeyer was in the lab as usual. She took out a fossil chip that had been in the acid for three days and put it under the microscope to take a picture. "[The chip] was curved so much, I couldn't get it in focus," Wittmeyer recalls. She used forceps to flatten it. "My forceps kind of sunk into it, made a little indentation, and it curled back up. I was like, stop it!" Finally, through her irritation, she realized what she had: a fragment of dinosaur soft tissue left behind when the mineral bone around it had dissolved. Suddenly Schweitzer and Wittmeyer were dealing with something no one else had ever seen. For a couple of weeks, Wittmeyer said, it was like Christmas every day.

In the lab, Wittmeyer now takes out a dish with six compartments, each holding a little brown dab of tissue in clear liquid, and puts it under the microscope lens. Inside each specimen is a fine network of almost-clear branching vessels — the tissue of a female *Tyrannosaurus rex* that strode through the forests 68 million years ago, preparing to lay eggs. Close up, the blood vessels from that *T. rex* and her ostrich cousins look remarkably alike. Inside the dinosaur vessels are things Schweitzer diplomatically calls "round microstructures" in the journal article, out of an abundance of scientific caution, but they are red and round, and she and other scientists suspect that they are red blood cells.

Of course, what everyone wants to know is whether DNA might be lurking in that tissue. Wittmeyer, from much experience with the press since the discovery, calls this "the awful question" — whether Schweitzer's work is paving the road to a real-life version of science fiction's *Jurassic Park,* where dinosaurs were regenerated from DNA preserved in amber. But DNA, which carries the genetic script for an animal, is a very fragile molecule. It's also ridiculously hard to study because it is so easily contaminated with modern biological material, such as microbes or skin cells, while buried or after being dug up. Instead, Schweitzer has been testing her dinosaur tissue samples for proteins, which are a bit hardier and more readily distinguished from contaminants. Specifically, she's been looking for collagen, elastin, and hemoglobin. Collagen makes up much of the bone scaffolding, elastin is wrapped around blood vessels, and hemoglobin carries oxygen inside red blood cells.

Because the chemical makeup of proteins changes through evolution, scientists can study protein sequences to learn more about how dinosaurs evolved. And because proteins do all the work in the body, studying them could someday help scientists understand dinosaur physiology — how their muscles and blood vessels worked, for example.

Proteins are much too tiny to pick out with a microscope. To look for them, Schweitzer uses antibodies, immune system molecules that recognize and bind to specific sections of proteins. Schweitzer and Wittmeyer have been using antibodies to chicken collagen, cow elastin, and ostrich hemoglobin to search for similar molecules in the dinosaur tissue. At an October 2005 paleontology conference, Schweitzer presented preliminary evidence that she has detected real dinosaur proteins in her specimens.

Further discoveries in the past year have shown that the discovery of soft tissue in B. rex wasn't just a fluke. Schweitzer and Wittmeyer have now found probable blood vessels, bone-building cells, and connective tissue in another *T. rex,* in a theropod from Argentina, and in a 300,000-year-old woolly mammoth fossil. Schweitzer's work is "showing us we really don't understand decay," Holtz says. "There's a lot of really basic stuff in nature that people just make assumptions about."

Young-earth creationists also see Schweitzer's work as revolutionary, but in an entirely different way. They first seized upon Schweitzer's work after she wrote an article for the popular science magazine *Earth* in 1997 about possible red blood cells in her dinosaur specimens. *Creation* magazine claimed that Schweitzer's research was "powerful testimony against the whole idea of dinosaurs living millions of years ago. It speaks volumes for the Bible's account of a recent creation."

This drives Schweitzer crazy. Geologists have established that the Hell Creek Formation, where B. rex was found, is 68 million years old, and so are the bones buried in it. She's horrified that some Christians accuse her of hiding the true meaning of her data. "They treat you really bad," she says. "They twist your words and they manipulate your data." For her, science and religion represent two different ways of looking at the world; invoking the hand of God to explain natural phenomena breaks the rules of science. After all, she says, what God asks is faith, not evidence. "If you have all this evidence and proof positive that God exists, you don't need faith. I think he kind of designed it so that we'd never be able to prove his existence. And I think that's really cool."

By definition, there is a lot that scientists don't know, because the whole point of science is to explore the unknown. By being clear that scientists haven't explained everything, Schweitzer leaves room for other explanations. "I think that we're always wise to leave certain doors open," she says.

But Schweitzer's interest in the long-term preservation of molecules and cells does have an otherworldly dimension: she's collaborating with NASA scientists on the search for evidence of possible past life on Mars, Saturn's moon Titan, and other heavenly bodies. (Scientists announced this spring, for instance, that Saturn's tiny moon Enceladus appears to have liquid water, a probable precondition for life.) Astrobiology is one of the wackier branches of biology, dealing in life that might or might not exist and might or might not take any recognizable form. "For almost everybody who works on NASA stuff, they are just in hog heaven, working on astrobiology questions," Schweitzer says. Her NASA research involves using antibodies to probe for signs of life in unexpected places. "For me, it's the means to an end. I really want to know about my dinosaurs."

To that purpose, Schweitzer, with Wittmeyer, spends hours in front of microscopes in dark rooms. To a fourth-generation Montanan, even the relatively laid-back Raleigh area is a big city. She reminisces wistfully about scouting for field sites on horseback in Montana. "Paleontology by microscope is not that fun," she says. "I'd much rather be out tromping around."

"My eyeballs are just absolutely fried," Schweitzer says after hours of gazing through the microscope's eyepieces at glowing vessels and blobs. You could call it the price she pays for not being typical.

PATRICIA GADSBY

Cooking for Eggheads

FROM *Discover*

PARIS IS SWELTERING, freakishly hot for an early June morning, and like much of the old city, the lab occupied by Hervé This at the Collège de France, a stone's throw from the venerable Sorbonne, has no air conditioning. As usual, however, This — pronounced "tiss" — looks dapper in a black suit and one of the impeccable white collarless shirts that have become his trademark. A full day lies ahead in his lab, he says, but first we must shop. He bounds to his feet, ditches his jacket, and descends to the stifling street below, proceeds down a cobbled alley, crosses the Boulevard Saint-Michel, rounds a corner, and dives into the local *supermarché*. He emerges with two dozen eggs and a cold brick of Normandy butter, his face crinkling into a grin. "For our experiments!" he announces. He has yet to break a sweat.

This is head of the molecular gastronomy group in the Collège de France's Laboratory for the Chemistry of Molecular Interactions. That's a mouthful to describe a lab that studies something simple: how the process of cooking changes the structure and taste of food. Nonetheless, molecular gastronomy marks the cutting edge of epicurism these days. Anyone who wields a saucepan is doing chemistry and physics, yet how many of us actually know what's going on in there? Molecular gastronomy aims to apply the piercing clarity of science to the culinary arts. Already in France, which takes the pleasures of the table seriously, molecular gastronomy is an officially recognized, government-funded science.

"Why molecular gastronomy?" asks This, heading off a question he's been asked many times before. "It sounds a little pompous, no? Why not . . . molecular cooking?" Easy, he replies. Cooking aims to produce a dish; it is a craft, a technique. Gastronomy is knowledge, albeit knowledge that can improve your cooking and your appreciation of it. Gastronomy is the science of anything to do with human nourishment, says This, more or less quoting Jean-Anthelme Brillat-Savarin, France's great food philosopher. Writing in 1825, Brillat-Savarin envisaged a discipline that would meld the physics and chemistry of food and cookery with the physiology of eating and especially with the glorious, sensual world of taste.

The "molecular" preface was added in the late 1980s by This and his late colleague, Nicholas Kurti, to evoke the chemical units that make up the water, fats, carbohydrates, proteins, and other compounds in food. Molecular had a dynamic, modern ring to it, perfect for ushering gastronomy into a new era. Besides, molecular gastronomy sounds so much more fun, sophisticated, and cultured than plain old "food science," a field with which it somewhat overlaps but that is largely geared to the mass-market needs of the food industry.

Not that This is a patronizing food snob. (Snobbism would be incompatible with his quest for objectivity, after all.) He would wholeheartedly agree with Brillat-Savarin that "a humble boiled egg" is as worthy of attention as "the banquets of kings." "If all you have to eat is this," he says, plucking an egg from its box and holding it between his thumb and forefinger, "it's important to cook it well."

Do we, though? The standard way to hard-boil eggs in Europe and America — ten minutes in boiling water — is not ideal, says This. The trouble, he notes clinically, is that 212 degrees Fahrenheit is far higher than the temperature at which the egg whites and the yolks coagulate. Egg whites are made up of protein and water (yolks contain fat as well). As eggs cook, their balled-up proteins uncoil into strands, and the strands bind together to form an intricate mesh that traps water. In essence, the proteins form a gel, a liquid dispersed in a solid. Boiling causes too many egg proteins to bind and form dense meshes, "so there is less sensation of water in the mouth," says This. Voilà: rubbery egg whites and sandy, grayish yolks.

The ten-minute egg is just the start of kitchen dogma. Our cookbooks are full of tips, caveats, and stipulations — *précisions,* as This calls them — drawn untested from tradition and folklore. "Cook meat at high temperature to seal in the juices? We've done the test — it's not true," says This. Use only eggs at room temperature for making mayonnaise? Not true either. Season steak with salt before cooking, or salt it afterward? Makes no difference, as the salt doesn't penetrate the meat. Parsing French recipes for a quarter-century in his quest for gastronomic clarity, This has identified more than 25,000 such admonitions; so far only a few hundred have been investigated. So many précisions, so little time.

This, a physical chemist and a former editor of the magazine *Pour La Science,* first began his testing as a sideline, alone in a laboratory he'd set up at home. Then he met Kurti, the man who would become his colleague and friend. Kurti was a low-temperature physicist at Oxford University and an irrepressible bon vivant. If there is a father of molecular gastronomy, Kurti is he. Thirty-five years ago, he was already poking the probe of a thermocouple into a cheese soufflé to take its internal temperature, the better to track its vapor-assisted ascent. "We know better the temperature inside the stars than inside a soufflé," Kurti once lamented.

They must have made an odd couple: the short, rotund, Hungarian-born Kurti and the tall, dashing, much younger This. Together they formed the International Workshops on Molecular Gastronomy and began corralling colleagues keen on kitchen science: the American food scholar Harold McGee and the British physicist Peter Barham along with open-minded chefs, critics, and writers who were passionate about food and good-humored enough to put their dearly held ideas (not to mention their egos) to pitiless scientific test. As a meeting place they chose Erice, a monastic town on a Sicilian mountaintop that was already a favorite retreat for physicists like Kurti. Although Kurti died in 1998, the motley group continues to meet every few years to trade information, ideas, and occasional insults, share a few late-night glasses of the local marsala, improvise a test kitchen in a monastery courtyard, and form the foundations for a truly modern cuisine.

The workshops are dizzying affairs. Topics for the texture workshop five years ago included the biomechanics of chewing and swallowing; the structure of meat and how cooking affects it; foams and gels, featuring custards and chocolate mousses; the effects of microwaves on spongy foodstuffs like eggplants and mushrooms;

and, on the last afternoon, a marathon session on the fractal nature of baba au rhum dough, conducted by a group of cantankerous physicists. "Well, that's why they are workshops, not lectures — to encourage the free exchange of ideas," This commented at the end of the day.

One participant at the texture workshop was Heston Blumenthal, a radical young British chef and the owner of the Fat Duck, near Windsor. Blumenthal was already receiving raves for the melting tenderness he coaxes out of lamb, achieved through his understanding of how heat diffuses in meat, and for the creation of a fabulous cookie — one that fizzes carbon dioxide in your mouth like so many tiny champagne bubbles. Today the Fat Duck has three Michelin stars and a biochemistry grad student in its development kitchen. Blumenthal is increasingly mentioned in the same breath as Ferran Adrià, the legendary Catalan chef at El Bulli, in Roses, two hours from Barcelona, whose superinventive and rather cerebral cuisine has drawn inspiration from the laboratory for years.

The ascent of the nerdy chef in Europe hasn't gone unnoticed in the United States. Suddenly science — once regarded with suspicion by foodies — looks like the next new thing. The term *molecular gastronomy* has begun popping up in restaurant reviews and on the food blog eGullet as a label for any edgy, out-there cuisine that combines unusual ingredients and employs techie gadgets.

"Because the phrase was around and catchy, it got applied to anyone experimenting with food," says McGee, an Erice regular. "Let's just say that many people aren't using molecular gastronomy to mean what Hervé means by it."

The confounding of molecular gastronomy with a sort of hipster cuisine drives even a patient man like This a little crazy. *Non, non, non:* molecular gastronomy isn't a cooking style, he insists. "We shouldn't confuse science with technology. Molecular gastronomy is only the science part. It asks, How does something work? What is the mechanism? The application of that knowledge is the cooking part, and that's technology. Cooking is a technique" — his voice softens — "combined with art." He adds, "Here in the lab, we do the science part — experiments." He introduces a Spanish student whose doctoral thesis investigates the effect of heat on two vegetable pigments, chlorophyll and carotenoids. Other students, using chromatography and nuclear magnetic resonance spectroscopy, for instance, are studying complex mixtures like meat and vegetable stocks. "You will see, we will do some experiments too," This says. "They will be simple, don't worry." You really can try this at home.

We begin by tackling the "standard model," the ten-minute egg. Can it be improved upon? Well, says This, if your grandmother cooked eggs that way for you, and you adored her and her cooking, there'll be no persuading you of a better way. (As This is fond of saying, "The most important ingredient in cooking is love.") But if you're willing to learn a little egg-protein chemistry, you can calibrate your eggs with astonishing exactitude.

Recall that when an egg cooks, its proteins first unwind and then link to form a rigidifying mesh. But not all its proteins solidify at the same temperature. Ovotransferrin, the first of the egg-white proteins to uncoil, begins to set at around 61 degrees Celsius, or 142°F. Ovalbumin, the most abundant egg-white protein, coagulates at 184°F. Yolk proteins generally fall in between, with most starting to solidify when they approach 158°F. Thus, cooking an egg at 158°F or so should achieve both a firmed-up yolk and still-tender whites, since at that low temperature only some of the egg-white proteins will have coagulated.

"Cooking eggs is really a question of temperature, not time," says This. To make the point, he switches on a small oven, sets the thermostat at 65°C, or 149°F, takes four eggs straight from the box, and unceremoniously places them inside. "I use an oven in the lab; it's easier. But if the oven in your kitchen is not accurate, cook eggs in plenty of water, using a good thermometer." About an hour later — timing isn't critical, and the eggs can stay in the oven for hours or even overnight — he retrieves the first egg and carefully shells it. "The sixty-five-degree egg!" he announces. The egg is unlike any I've eaten. The white is as delicately set and smooth as custard, and the yolk is still orange and soft. It's not hard to see why *l'oeuf à soixante-cinq degrés* is becoming the rage with chefs in France. (Salmonella can't survive more than a few minutes at 60°C, or 140°F, so a 65-degree egg cooked for an hour should be quite safe.)

Next This turns up the oven thermostat to 67°C, or 153°F, and after waiting a while for the eggs inside to reach that temperature — again, he's casual about the timing — he retrieves a second one:

"The sixty-seven-degree egg!" At this temperature the yolk has just started thickening up — some of its proteins have coagulated, but the majority have not. "Look, you can mold it," he says, scooping out the yolk and manipulating the pliable orangey-yellow ball like fresh Play-Doh. He tries to mold a heart, then settles for a cube.

"Try one," he says, taking a third egg from the oven for me to play with before turning up the heat to 158°F (70°C). The 70-degree egg, when it is finally done, has a moistly set yolk and a very tender white. "So you see, you can adjust the temperature depending on what you want," says This. If you prefer a firmer egg, cook it at 167°F or 176°F. Bear in mind, though, that the most copious of the egg-white proteins sets at 184°F — hence the rubbery results of the 212-degree bath.

So familiar is This with this process that he can tell at a glance the temperature at which an egg was cooked. During lunch at a local bistro, I notice a 65-degree-Celsius egg on the menu, served on a *fricassée de girolles.* As the plate is set down, This says, "That's not a sixty-five-degree egg. It's a sixty-four-degree egg." The yolk is soft, and the egg white, while completely opaque, is so delicately jelled and fragile that it breaks apart slightly when it is plated. "Eh, *oui,*" the chef sighs; he is having *des ennuis* regulating the heat of his stove. Never mind that the presentation isn't completely perfect — the egg, mixing in with the earthy mushroom stew, is delicious.

Back in his laboratory, This puts on a lab coat to protect his shirt "so my wife won't complain that I make spots on it." He breaks a raw egg one-handed, plops the white into a bowl, and starts rapidly whisking. Whisking, of course, incorporates air into the aqueous white. It also causes some proteins in the egg white to unfold. The resulting protein strands then form a mesh around the air bubbles, stabilizing the foam. Usually an egg white produces about half a pint of foam.

Why not more? asks This, whisking like a demon. It can't be lack of air — there's an endless supply — so it must be lack of water. He adds a squirt to the beaten egg white, whisks again, squirts water, whisks some more. The snowy mass keeps growing. "If I went on beating I could get liters and liters — gallons of foam! — from *one* egg white," he says, pausing at last to wipe his brow. With a thunderstorm brewing, the lab feels muggier than ever, and even This shows signs of wanting a break. "So you see, you only need one egg to make a lot of mousse, enough for a dinner party." However, the foam will be less stable, because the viscosity of that single egg white has been diluted. Whisking hard helps, as smaller bubbles are more stable. So does beating in sugar (or gum), which stabilizes the foam by increasing its viscosity.

The voluminous-egg-white stunt was first used for an educational project in French schools. Later it occurred to This that it could be put to culinary use. Eight years ago he struck up a collaboration with one of France's most lionized chefs, Pierre Gagnaire. The two regularly rendezvous to brainstorm: This tosses up new culinary concepts based on his scientific musings, and Gagnaire transforms them into elegant recipes. This proposed replacing water in the expanded egg-white foam with a flavorful liquid to make an ethereal perfumed meringue, an invention he named *cristaux de vent,* or "wind crystals." Gagnaire's creation: a soft black olive buried inside a crisp meringue made light as air with the olive's own pickling brine.

This and Kurti envisaged a day when molecular gastronomy would help people cook in entirely different ways; they never guessed that day might come so soon. Quite a few chefs are toying with chemicals pulled from the lab shelf, using new jelling agents, for instance, to encapsulate sauces and liquids in a fragile, jellified skin, like salmon eggs. "It's truly the terra incognita of cooking," This says: full of potential for brilliant, thrilling innovation, or for dreadful mischief if explored without discernment. And it's bound to stir up a stew of admiration, bemusement, wonder, and ridicule — in short, gloriously animated debate.

The following day This has an appointment with Gagnaire at his restaurant on a corner of the Rue Balzac. You could call their collaboration a tech transfer. "Mad," says This affectionately of his collaborator in the taxi on the way there. "Even madder than me."

Gagnaire is a little late, fresh off the Eurostar from a conference in London. He has a world-weary mien, or maybe he's just tired: intense pale blue eyes, an angular nose, a slight beard, flopping hair — a corsair in chef's whites. The two men embrace happily on meeting and quickly settle down to business in the salon adjoining the dining room. This's laptop comes out, and notebooks appear on the table. This begins serving up a generous helping of ideas, his mind racing; Gagnaire keeps up, peppering him with questions, taking notes, riffing on This's suggestions. Yes, yes, the texture of a taste is very important, says Gagnaire during a discussion of various jelling agents like methylcellulose and alginates. A taste can change with a change in the texture or the surface of the food. Which scientist, Gagnaire wants to know, has studied the effect of surfaces on taste? Later This reviews plans for an upcoming dinner in which each dish will be named for a scientist and evoke his work: *le dessert* Einstein, *le plat* Faraday, *la sauce* Pasteur.

"Okay, *d'accord,*" Gagnaire says gallantly. Jazz plays softly over the restaurant's sound system. There's a clink of cutlery, a murmur of appreciative conversation from the dining room. Fleets of beautiful little dishes begin to arrive. One, a dish of sweet mild scallops and mussels in a pool of consommé, is enlivened by a rust-colored cream of *araignée de mer* — spider crab — that floods the mouth and nose with sea aromas. Beside it sits a salad of meaty tomatoes, with tangled greens and seaweed on a sauce dark as night, and white wind crystals, the ethereal meringues that Gagnaire impregnates with black olives. It's a deceptively simple combination: plush tomatoes, the tang of slippery seaweed, the meringue crisp and sweetish at first bite, then soft and pungent with olive on the inside.

As soon as this dish is whisked away, I want more. Jean-Anthelme Brillat-Savarin, the original gastronome, was on to something. "The creation of a new dish," he wrote in *The Physiology of Taste,* "does more for the happiness of mankind than the discovery of a new star." If Nicholas Kurti is the father of molecular gastronomy and Hervé This is its son, then Brillat-Savarin surely is its holy ghost.

JAMES GLEICK

Cyber-Neologoliferation

FROM *The New York Times Magazine*

WHEN I GOT to John Simpson and his band of lexicographers in Oxford earlier this fall, they were working on the P's. *Pletzel, plish, pod person, point-and-shoot, polyamorous* — these words were all new, one way or another. They had been plowing through the P's for two years but were almost done (except that they'll never be done), and the Q's will be "just a twinkle of an eye," Simpson said. He prizes patience and the long view. A pale, soft-spoken man of middle height and profound intellect, he is chief editor of the *Oxford English Dictionary* and sees himself as a steward of tradition dating back a century and a half. "Basically it's the same work as they used to do in the nineteenth century," he said. "When I started in 1976, we were still working very much on these index cards, everything was done on these index cards." He picked up a stack of 6-inch-by-4-inch slips and riffled through them. A thousand of these slips were sitting on his desk, and within a stone's throw were millions more, filling metal files and wooden boxes with the ink of two centuries, words, words, words.

But the word slips have gone obsolete now, as Simpson well knows. They are treeware (a word that entered the *OED* in September as "computing slang, freq. humorous"). *Blog* was recognized in 2003, *dot-commer* in 2004, *metrosexual* in 2005, and the verb *Google* last June. Simpson has become a frequent and accomplished Googler himself, and his workstation connects to a vast and interlocking set of searchable databases, a better and better approximation of what might be called All Previous Text. The *OED* has met the Internet, and however much Simpson loves the *OED*'s roots and legacy, he is leading a revolution, willy-nilly — in what it is, what it knows, what it sees.

The English language, spoken by as many as 2 billion people in every country on earth, has entered a period of ferment, and this place may be the best observation platform available. The perspective here is both intimate and sweeping. In its early days, the *OED* found words almost exclusively in books; it was a record of the formal written language. No longer. The language upon which the lexicographers eavesdrop is larger, wilder, and more amorphous; it is a great, swirling, expanding cloud of messaging and speech: newspapers, magazines, pamphlets; menus and business memos; Internet news groups and chat-room conversations; and television and radio broadcasts.

The *OED* is unlike any other dictionary, in any language. Not simply because it is the biggest and the best, though it is. Not just because it is the supreme authority. (It wears that role reluctantly: it does not presume, or deign, to say that any particular usage or spelling is correct or incorrect; it aims merely to capture the language people use.) No, what makes the *OED* unique is a quality for which it can only strive: completeness. It wants every word, all the lingo: idioms and euphemisms, sacred or profane, dead or alive, the King's English or the street's. The *OED* is meant to be a perfect record, perfect repository, perfect mirror of the entire language.

James Murray, the editor who assembled the first edition through the final decades of the nineteenth century, was really speaking of the *language* when he said, in 1900, "The English Dictionary, like the English Constitution, is the creation of no one man, and of no one age; it is a growth that has slowly developed itself adown the ages." And developing faster nowadays. The *OED* tries to grasp the whole arc of an ever-changing history. Murray knew that with "adown" he was using a word that could be dated back to Anglo-Saxon of the year 975. When John Updike begins his *New Yorker* review of the new John le Carré novel by saying, "Hugger-mugger is part of life," it is the *OED* that gives us the first recorded use of the word, in 1529 (". . . not alwaye whyspered in hukermoker," Sir Thomas More) and twenty-seven more quotations from four different centuries. But when the *New York Times* prints a timely editorial about "sock puppets," meaning false identities assumed on the Internet, the *OED* has more work to do.

The version now under way is only the third edition. The first, containing 414,825 words in ten weighty volumes, was presented to King George V and President Coolidge in 1928. Several "supplements" followed, but not till 1989 did the second edition appear: twenty volumes, totaling 21,730 pages. It weighed 138 pounds.

The third edition is a mutation. It is weightless, taking its shape in the digital realm. To keyboard it, Oxford hired a team of 150 typists in Florida for eighteen months. (That was before the verb *keyboard* had even found its way in, as Simpson points out, not to mention the verb *outsource*.) No one can say for sure whether *OED3* will ever be published in paper and ink. By the point of decision, not before twenty years or so, it will have doubled in size yet again. In the meantime, it is materializing before the world's eyes, bit by bit, online. It is a thoroughgoing revision of the entire text. Whereas the second edition just added new words and new usages to the original entries, the current project is researching and revising from scratch — preserving the history but aiming at a more coherent whole.

The revised installments began to appear online in the year 2000. Simpson chose to begin the revisions not with the letter A but with M. Why? It seems the original *OED* was not quite a seamless masterpiece. Murray did start at A, logically, and the early letters show signs of the enterprise's immaturity. The entries in A tended to be smaller, with different senses of a word crammed together instead of teased lovingly apart in subentries. "It just took them a long time to sort out their policy and things," Simpson says, "so if we started at A, then we'd be making our job doubly difficult. I think they'd sorted themselves out by . . ." He stops to think. "Well, I was going to say D, but Murray always said that E was the worst letter, because his assistant, Henry Bradley, started E, and Murray always said that he did that rather badly. So then we thought, Maybe it's safe to start with G, H. But you get to G and H, and there's I, J, K, and you know, you think, well, start after that."

So the first wave of revision encompassed 1,000 entries from *M* to *mahurat*. The rest of the M's, the N's, and the O's have followed in due course. That's why, at the end of 2006, John Simpson and his lexicographers are working on the P's. Their latest quarterly installment, in September, covers *pleb* to *Pomak*. Simpson mentions rather proudly that they scrambled at the last instant to update the entry for *Pluto* when the International Astronomical Union voted to rescind its planethood. *Pluto* had entered the second edition as "1. A small planet of the solar system . . ." discovered in 1930 and "2. The name of a cartoon dog . . ." first appearing in 1931. The Disney meaning was more stable, it turns out. In *OED3,* Pluto is still a dog but merely "a small planetary body."

Even as they revise the existing dictionary in sequence, the *OED* lexicographers are adding new words wherever they find them, at an accelerating pace. Beside the P's, September's freshman class included *agroterrorism, bahookie* (a body part), *beer pong* (a drinking game), *bippy* (as in, you bet your —), *chucklesome, cypherpunk, tuneage,* and *wonky.* Every one of these underwent intense scrutiny. The addition of a new word is a solemn matter.

"Because it's the *OED,*" says Fiona McPherson, a new-words editor, "once something goes in, it cannot ever come out again." In this respect, you could say that the *OED* is a *roach motel* (added March 2005: "Something from which it may be difficult or impossible to be extricated"). A word can go *obs.* or *rare,* but the editors feel that even the most ancient and forgotten words have a way of coming back — people rediscover them or reinvent them — and anyway, they are part of the language's history.

The new-words department, where that history rolls forward, is not to everyone's taste. "I love it, I really really love it," McPherson says. "You're at the cutting edge, you're dealing with stuff that's not there and you're, I suppose, shaping the language. A lot of people are more interested in the older stuff; they like nothing better than reading through eighteenth-century texts looking for the right word. That doesn't suit me as much, I have to say." *Cutting edge,* incidentally, is not a new word: according to the *OED,* H. G. Wells used it in its modern sense in 1916.

As a rule, a neologism needs five years of solid evidence for admission to the canon. "We need to be sure that a word has established a reasonable amount of longevity," McPherson says. "Some things do stick around that you would never expect to stick around, and then other things, you think *that* will definitely be around, and everybody talks about it for six months, and then . . ."

Still, a new word as of September is *bada-bing:* American slang "suggesting something happening suddenly, emphatically, or easily and predictably." *The Sopranos* gets no credit. The historical citations begin with a 1965 audio recording of a comedy routine by Pat Cooper and continue with newspaper clippings, a television news transcript, and a line of dialogue from the first *Godfather* movie: "You've gotta get up close like this and bada-bing! you blow their brains all over your nice Ivy League suit." The lexicographers also provide an etymology, a characteristically exquisite piece of guesswork: "Origin uncertain. Perh. imitative of the sound of a drum roll and cymbal clash. . . . Perh. cf. Italian bada bene mark well."

But is *bada-bing* really an official part of the English language? What makes it a word? I can't help wondering, when it comes down to it, isn't *bada-bing* (also *badda-bing, badda badda bing, badabing, badaboom*) just a noise?

"I dare say the thought occurs to editors from time to time," Simpson says. "But from a lexicographical point of view, we're interested in the conventionalized representation of strings that carry meaning. Why, for example, do we say *Wow!* rather than some other string of letters? Or *Zap!* Researching these takes us into interesting areas of comic-magazine and radio-TV-film history and other related historical fields. And it often turns out that they became institutionalized far earlier than people nowadays may think."

When Murray began work on *OED1,* no one had any idea how many words were there to be found. Probably the best and most comprehensive dictionary of English was American, Noah Webster's: 70,000 words. That number was a baseline. Where were the words to be discovered? For the first editors it went almost without saying that the source, the wellspring, should be the *literature* of the language. Thus it began as a dictionary of the written language, not the spoken language. The dictionary's first readers combed Milton and Shakespeare (still the single most quoted author, with more than 30,000 references), Fielding and Swift, histories and sermons, philosophers and poets. "A thousand readers are wanted," Murray announced in his famous 1879 public appeal. "The later sixteenth-century literature is very fairly done; yet here several books remain to be read. The seventeenth century, with so many more writers, naturally shows still more unexplored territory." He considered the territory to be large, but ultimately finite.

It no longer seems finite.

"We're painting the Forth Bridge!" says Bernadette Paton, an associate editor. "We're running the wrong way on a travolator!" (I get the first part — "allusion to the huge task of maintaining the painted surfaces of the railway bridge over the Firth of Forth" — but I have to ask about *travolator*. Apparently it's a moving sidewalk.)

The *OED* is a historical dictionary, providing citations meant to show the evolution of every word, beginning with the earliest known usage. So a key task, and a popular sport for thousands of volunteer word aficionados, is antedating: finding earlier citations than those already known. This used to be painstakingly slow and chancy. When Paton started in new words, she found herself struggling with *headcase*. She had current citations, but she says she felt sure it must be older, and books were of little use. She wandered around the office muttering *headcase, headcase, headcase*. Suddenly one of her colleagues started singing: "My name is Bill, and I'm a *headcase* / They practice making up on my face." She perked up.

"What date would that be?" she asked.

"I don't know, it's a Who song," he said, "1966 probably, something like that."

So "I'm a Boy," by P. Townshend, became the *OED*'s earliest citation for *headcase*.

Antedating is entirely different now: online databases have opened the floodgates. Lately Paton has been looking at words starting with *pseudo-*. Searching through databases of old newspapers and historical documents has changed her view of them. "I tended to think of *pseudo-* as a prefix that just took off in the sixties and seventies, but now we find that a lot of them go back much earlier than we thought." Also in the P's, *poison pen* has just been antedated with a 1911 headline in the *Evening Post* in Frederick, Maryland. "You get the sense that this sort of language seeps into local newspapers first," she says. "We would never in a million years have sent a reader to read a small newspaper like that."

The job of a new-words editor felt very different precyberspace, Paton says: "New words weren't proliferating at quite the rate they have done in the last ten years. Not just the Internet, but text messaging and so on has created lots and lots of new vocabulary." Much of the new vocabulary appears online long before it will make it into books. Take *geek*. It was not till 2003 that *OED3* caught up with the main modern sense: "a person who is extremely devoted to and knowledgeable about computers or related technology." Internet chitchat provides the earliest known reference, a posting to a Usenet newsgroup, net.jokes, on February 20, 1984.

The scouring of the Internet for evidence — the use of cyberspace as a language lab — is being systematized in a program called the Oxford English Corpus. This is a giant body of text that begins in 2000 and now contains more than 1.5 billion words, from published material but also from Web sites, Weblogs, chat rooms, fanzines, corporate home pages, and radio transcripts. The corpus sends its home-built Web crawler out in search of text, raw material to show how the language is really used.

I'm too embarrassed to ask the lexicographers if they have a favorite word. They get that a lot. Peter Gilliver tells me his anyway: *twiffler.* A twiffler, in case you didn't know, is a plate intermediate in size between a dinner plate and a bread plate. "I love it because it fills a gap," Gilliver says. "I also love it because of its etymology. It comes from Dutch, like a lot of ceramics vocabulary. *Twijfelaar* means something intermediate in size, and it comes from *twijfelen,* which means to be unsure. It's a plate that can't make up its mind!"

Fiona McPherson gives me *mondegreen.* A mondegreen is a misheard lyric, as in "Lead on, O kinky turtle." It is named after Lady Mondegreen. There was no Lady Mondegreen. The lines of a ballad, "They hae slain the Earl of Murray, / And laid him on the green" are misheard as "They have slain the Earl of Murray and Lady Mondegreen."

"A lot of people are just really excited by that word because they think it's amazing that there is a word for that concept," McPherson says.

I have my own favorites among the newest entries in *OED3. Pixie dust* is, as any child knows, "an imaginary magical substance used by pixies." *Air kiss* is defined with careful anatomical instructions plus a note: "sometimes with the connotation that such a gesture implies insincerity or affectation." *Builder's bum* is reportedly Brit. and colloq., "with allusion to the perceived propensity of builders to expose inadvertently this part of the body."

It is clear that the English of the *OED* is no longer the purely written language, much less a formal or respectable English, the diction recommended by any authority. Gilliver, a longtime editor who also seems to be the *OED*'s resident historian, points out that the dictionary feels obliged to include words that many would regard simply as misspellings. No one is particularly proud of the new entry as of December 2003 for *nucular,* a word not associated with high standards of diction. "Bizarrely, I was amazed to find that the spelling n-u-c-u-l-a-r has decades of history," Gilliver says. "And that is not to be confused with the quite different word *nucular,* meaning 'of or relating to a nucule.'" There is even a new entry for *miniscule;* it has citations going back more than one hundred years.

Yet the very notion of correct and incorrect spelling seems under attack. In Shakespeare's day there was no such thing: no right and wrong in spelling, no dictionaries to consult. The word *debt* could be spelled det, dete, dett, dette, or dept, and no one would complain.

Then spelling crystallized, with the spread of printing. Now, with mass communication taking another leap forward, spelling may be diversifying again, spellcheckers notwithstanding. The *OED* so far does not recognize *straight-laced,* but the Oxford English Corpus finds it outnumbering *strait-laced.* Similarly for *just desserts.*

To explain why cyberspace is a challenge for the *OED* as well as a godsend, Gilliver uses the phrase "sensitive ears."

"You know we are listening to the language," he says. "When you are listening to the language by collecting pieces of paper, that's fine, but now it's as if we can hear everything said anywhere. Members of some tiny English-speaking community anywhere in the world just happen to commit their communications to the Web: there it is. You thought some word was obsolete? Actually, no, it still survives in a very small community of people who happen to use the Web — we can hear about it."

In part, it's just a problem of too much information: a small number of lexicographers with limited time. But it's also that the *OED* is coming face to face with the language's boundlessness.

The universe of human discourse always has backwaters. The language spoken in one valley was a little different from the language of the next valley and so on. There are more valleys now than ever, but they are not so isolated. They find one another in chat rooms and on blogs. When they coin a word, anyone may hear.

Neologisms can be formed by committee: *transistor,* Bell Laboratories, 1948. Or by wags: *booboisie,* H. L. Mencken, 1922. But most arise through spontaneous generation, organisms appearing in a petri dish, like *blog* (c. 1999). If there is an ultimate limit to the sensitivity of lexicographers' ears, no one has yet found it. The rate of change in the language itself — particularly the process of neologism — has surely shifted into a higher gear now, but away from dictionaries, scholars of language have no clear way to measure the process. When they need quantification, they look to the dictionaries.

"An awful lot of neologisms are spur-of-the-moment creations, whether it's literary effect or it's conversational effect," says Naomi S. Baron, a linguist at American University, who studies these issues. "I could probably count on the fingers of a hand and a half the serious linguists who know anything about the Internet. That hand and a half of us are fascinated to watch how the Internet makes it possible not just for new words to be coined but for neologisms to spread like wildfire."

It's partly a matter of sheer intensity. Cyberspace is an engine driving change in the language. "I think of it as a saucepan under which the temperature has been turned up," Gilliver says. "Any word, because of the interconnectedness of the English-speaking world, can spring from the backwater. And they are still backwaters, but they have this instant connection to ordinary, everyday discourse." Like the printing press, the telegraph, and the telephone before it, the Internet is transforming the language simply by transmitting information differently. And what makes cyberspace different from all previous information technologies is its intermixing of scales from the largest to the smallest without prejudice, broadcasting to the millions, narrowcasting to groups, instant messaging one to one.

So anyone can be an *OED* author now. And, by the way, many try. "What people love to do is send us words they've invented," Bernadette Paton says, guiding me through a windowless room used for storage of old word slips. *Will you put the word I have invented into one of your dictionaries?* is a question in the AskOxford.com FAQ. All the submissions go into the files, and until there is evidence for some general usage, that's where the wannabes remain.

Don't bother sending in *FAQ*. Don't bother sending in *wannabes*.

They're not even particularly new. For that matter, don't bother sending in anything you find via Google. "Please note," the *OED*'s Web site warns solemnly, "it is generally safe to assume that examples found by searching the Web, using search engines such as Google, will have already been considered by *OED* editors."

JOHN HORGAN

The Final Frontier

FROM *Discover*

ONE OF MY MOST MEMORABLE MOMENTS as a journalist occurred in December 1996, when I attended the Nobel Prize festivities in Stockholm. During a 1,300-person white-tie banquet presided over by Sweden's king and queen, David Lee of Cornell University, who shared that year's physics prize, decried the "doomsayers" claiming that science is ending. Reports of science's death "are greatly exaggerated," he said.

Lee was alluding to my book, *The End of Science,* released earlier that year. In it, I made the case that science — especially pure science, the grand quest to understand the universe and our place in it — might be reaching a cul-de-sac, yielding "no more great revelations or revolutions, but only incremental, diminishing returns." More than a dozen Nobel laureates denounced this proposition, mostly in the media but some to my face, as did the White House science adviser, the British science minister, the head of the Human Genome Project, and the editors in chief of the journals *Science* and *Nature.*

Over the past decade scientists have announced countless discoveries that seem to undercut my thesis: cloned mammals (starting with Dolly the sheep), a detailed map of the human genome, a computer that can beat the world champion in chess, brain chips that let paralyzed people control computers purely by thought, glimpses of planets around other stars, and detailed measurements of the afterglow of the Big Bang. Yet within these successes there are nagging hints that most of what lies ahead involves filling in the blanks of today's big scientific concepts, not uncovering totally new ones.

Even Lee acknowledges the challenge. "Fundamental discoveries are becoming more and more expensive and more difficult to achieve," he says. His own Nobel helps make the point. The Russian physicist Pyotr Kapitsa discovered the strange phenomenon known as superfluidity in liquid helium in 1938. Lee and two colleagues merely extended that work, showing that superfluidity also occurs in a helium isotope known as helium 3. In 2003, yet another Nobel Prize was awarded for investigations of superfluidity. Talk about anticlimactic!

Optimists insist that revolutionary discoveries surely lie just around the corner. Perhaps the big advance will spring from physicists' quest for a theory of everything; from studies of "emergent" phenomena with many moving parts, such as ecologies and economies; from advances in computers and mathematics; from nanotechnology, biotechnology, and other applied sciences; or from investigations of how brains make minds. "I can see problems ahead of all sizes, and clearly many of them are soluble," says physicist and Nobel laureate Philip Anderson (who, in 1999, coined the term *Horganism* to describe "the belief that the end of science . . . is at hand"). On the flip side, some skeptics contend that science can never end because all knowledge is provisional and subject to change.

For the tenth anniversary of *The End of Science,* I wanted to address these new objections. What I find is that the limits of scientific inquiry are more visible than ever. My goal, now as then, is not to demean valuable ongoing research but to challenge excessive faith in scientific progress. Scientists pursuing truth need a certain degree of faith in the ultimate knowability of the world; without it, they would not have come so far so fast. But those who deny any evidence that challenges their faith violate the scientific spirit. They also play into the hands of those who claim that "science itself is merely another kind of religion," as physicist Lawrence Krauss of Case Western Reserve University warns.

Argument: Predictions that science is ending are old hat, and they have always proved wrong. The most common response to *The End of Science* is the "that's what they thought then" claim. It goes like this: At the end of the nineteenth century, physicists thought they knew everything just before relativity and quantum mechanics blew physics wide open. Another popular anecdote involves a nineteenth-century U.S. patent official who quit his job because he thought "everything that can be invented has been invented." In fact, the patent-official story is purely apocryphal, and the description of nineteenth-century physicists as smug know-it-alls is greatly exaggerated. Moreover, even if scientists had foolishly predicted science's demise in the past, that does not mean all such predictions are equally foolish.

The "that's what they thought then" response implies that because science advanced rapidly over the past century or so, it must continue to do so, possibly forever. This is faulty inductive reasoning. A broader view of history suggests that the modern era of explosive progress is an anomaly — the product of a unique convergence of social, economic, and political factors — that must eventually end. Science itself tells us that there are limits to knowledge. Relativity theory prohibits travel or communication faster than light. Quantum mechanics and chaos theory constrain the precision with which we can make predictions. Evolutionary biology reminds us that we are animals, shaped by natural selection not for discovering deep truths of nature but for breeding.

The greatest barrier to future progress in science is its past success. Scientific discovery resembles the exploration of the Earth. The more we know about our planet, the less there is to explore. We have mapped out all the continents, oceans, mountain ranges, and rivers. Every now and then we stumble upon a new species of lemur in an obscure jungle or an exotic bacterium in a deep-sea vent, but at this point we are unlikely to discover something truly astonishing, like dinosaurs dwelling in a secluded cavern. In the same way, scientists are unlikely to discover anything surpassing the Big Bang, quantum mechanics, relativity, natural selection, or genetics.

Just over a century ago, the American historian Henry Adams observed that science accelerates through a positive feedback effect: knowledge begets more knowledge. This acceleration principle has an intriguing corollary. If science has limits, then it might be moving at maximum speed just before it hits the wall. I am not the only science journalist who suspects we have entered this endgame. "The questions scientists are tackling now are a lot narrower than those that were being asked one hundred years ago," Michael

Lemonick wrote in *Time* magazine recently, because "we've already made most of the fundamental discoveries."

Argument: Science is still confronting huge remaining mysteries, like where the universe came from. Other reporters like to point out that there is "No End of Mysteries," as a cover story in *U.S. News & World Report* put it. But some mysteries are probably unsolvable. The biggest mystery of all is the one cited by Stephen Hawking in *A Brief History of Time:* Why is there something rather than nothing? More specifically, what triggered the Big Bang, and why did the universe take this particular form rather than some other form that might not have allowed our existence?

Scientists' attempts to solve these mysteries often take the form of what I call ironic science — unconfirmable speculation more akin to philosophy or literature than genuine science. (The science is ironic in the sense that it should not be considered a literal statement of fact.) A prime example of this style of thinking is the anthropic principle, which holds that the universe must have the form we observe because otherwise we would not be here to observe it. The anthropic principle, championed by leading physicists such as Leonard Susskind of Stanford University, is cosmology's version of creationism.

Another example of ironic science is string theory, which for more than twenty years has been the leading contender for a "theory of everything" that explains all of nature's forces. The theory's concepts and jargon have evolved over the past decade, with two-dimensional membranes replacing one-dimensional strings, but the theory comes in so many versions that it predicts virtually everything — and hence nothing at all. Critics call this the "Alice's restaurant problem," a reference to a folk song with the refrain "You can get anything you want at Alice's restaurant." This problem leads Columbia mathematician Peter Woit to call string theory "not even wrong" in his influential blog of the same title, which refers to a famous put-down by Wolfgang Pauli.

Although Woit echoes the criticisms of string theory I made in *The End of Science,* he still hopes that new mathematical techniques may rejuvenate physics. I have my doubts. String theory already represents an attempt to understand nature through mathematical argumentation rather than empirical tests. To break out of its current impasse, physics desperately needs not new mathematics but new empirical findings — like the discovery in the late 1990s that the expansion of the universe is accelerating. This is by far the most exciting advance in physics and cosmology in the last decade, but it has not led to any theoretical breakthrough. Meanwhile, the public has become increasingly reluctant to pay for experiments that can push back the frontier of physics. The Large Hadron Collider will be the world's most powerful particle accelerator when it goes online next year, and yet it is many orders of magnitude too weak to probe directly the microrealm where strings supposedly dwell.

Argument: Science can't ever come to an end because theories, by their very nature, keep being overturned. Many philosophers — and a surprising number of scientists — accept this line, which essentially means that all science is ironic. They adhere to the postmodern position that we do not discover truth so much as we invent it; all our knowledge is therefore provisional and subject to change. This view can be traced back to two influential philosophers: Karl Popper, who held that theories can never be proved but only disproved, or falsified, and Thomas Kuhn, who contended that theories are not true statements about reality but only temporarily convenient suppositions, or paradigms.

If all our scientific knowledge were really this flimsy and provisional, then of course science could continue forever, with theories changing as often as fads in clothing or music. But the postmodern stance is clearly wrong. We have not invented atoms, elements, gravity, evolution, the double helix, viruses, and galaxies; we have discovered them, just as we discovered that Earth is round and not flat.

When I spoke to him more than ten years ago, philosopher Colin McGinn, now at the University of Miami, rejected the view that all of science is provisional, saying, "Some of it is, but some of it isn't!" He also suggested that, given the constraints of human cognition, science will eventually reach its limits; at that point, he suggests, "religion might start to appeal to people again." Today, McGinn stands by his assertion that science "must in principle be completable" but adds, "I don't, however, think that people will or should turn to religion if science comes to an end." Current events might suggest otherwise.

*

Argument: Reductionist science may be over, but a new kind of emergent science is just beginning. In his new book, *A Different Universe,* Robert Laughlin, a physicist and Nobel laureate at Stanford, concedes that science may in some ways have reached the "end of reductionism," which identifies the basic components and forces underpinning the physical realm. Nevertheless, he insists that scientists can discover profound new laws by investigating complex, emergent phenomena, which cannot be understood in terms of their individual components.

Physicist and software mogul Stephen Wolfram advances a similar argument from a more technological angle. He asserts that computer models called cellular automata represent the key to understanding all of nature's complexities, from quarks to economies. Wolfram found a wide audience for these ideas with his 1,200-page self-published opus *A New Kind of Science.* He asserts that his book has been seen as "initiating a paradigm shift of historic importance in science, with new implications emerging at an increasing rate every year."

Actually, Wolfram and Laughlin are recycling ideas propounded in the 1980s and 1990s in the fields of chaos and complexity, which I regard as a single field — I call it chaoplexity. Chaoplexologists harp on the fact that simple rules, when followed by a computer, can generate extremely complicated patterns, which appear to vary randomly as a function of time or scale. In the same way, they argue, simple rules must underlie many apparently noisy, complicated aspects of nature.

So far, chaoplexologists have failed to find any profound new scientific laws. I recently asked Philip Anderson, a veteran of this field, to list major new developments. In response he cited work on self-organized criticality, a mathematical model that dates back almost two decades and that has proved to have limited applications. One reason for chaoplexity's lack of progress may be the notorious butterfly effect, the notion that tiny changes in initial conditions can eventually yield huge consequences in a chaotic system; the classic example is that the beating of a butterfly's wings could eventually trigger the formation of a tornado. The butterfly effect limits both prediction and retrodiction, and hence explanation, because specific events cannot be ascribed to specific causes with complete certainty.

*

Argument: Applied physics could still deliver revolutionary breakthroughs, like fusion energy. Some pundits insist that although the quest to discover nature's basic laws might have ended, we are now embarking on a thrilling era of gaining control over these laws. "Every time I open the newspaper or search the Web," the physicist Michio Kaku of the City University of New York says, "I find more evidence that the foundations of science are largely over, and we are entering a new era of applications." Saying that science has ended because we understand nature's basic laws, Kaku contends, is like saying that chess ends once you understand the rules.

I see some validity to this point; that is why *The End of Science* focused on pure rather than applied research. But I disagree with techno-evangelists such as Kaku, Eric Drexler, and Ray Kurzweil that nanotechnology and other applied fields will soon allow us to manipulate the laws of nature in ways limited only by our imaginations. The history of science shows that basic knowledge does not always translate into the desired applications.

Nuclear fusion — a long-sought source of near-limitless energy and one of the key applications foreseen by Kaku — offers a prime example. In the 1930s, Hans Bethe and a handful of other physicists elucidated the basic rules governing fusion, the process that makes stars shine and thermonuclear bombs explode. Over the past fifty years, the United States alone has spent nearly $20 billion trying to control fusion in order to build a viable power plant. During that time, physicists repeatedly touted fusion as the energy source of the future.

The United States and other nations just agreed to invest another $13 billion to build the International Thermonuclear Experimental Reactor in France. Still, even optimists acknowledge that fusion energy faces formidable technical, economic, and political barriers all the same. William Parkins, a nuclear physicist and veteran of the Manhattan Project, recently advocated abandoning fusion-energy research, which he called "as expensive as it is discouraging." If there are breakthroughs here, the current generation probably will not live to see them.

Argument: We are on the verge of a breakthrough in applied biology that will allow people to live essentially forever. The potential applications of biology are certainly more exciting these days than those of physics. The completion of the Human Genome Project and recent advances in cloning, stem cells, and other fields have emboldened some scientists to predict that we will soon conquer not only disease but aging itself. "The first person to live to one thousand may have been born by 1945," declares computer-scientist-turned-gerontologist Aubrey de Grey, a leader in the immortality movement (who was born in 1963).

Many of de Grey's colleagues beg to differ, however. His view "commands no respect at all within the informed scientific community," twenty-eight senescence researchers declared in a 2005 journal article. Indeed, evolutionary biologists warn that immortality may be impossible to achieve because natural selection designed us to live only long enough to reproduce and raise our children. As a result, senescence does not result from any single cause or even a suite of causes; it is woven inextricably into the fabric of our bodies. The track record of two fields of medical research — gene therapy and the war on cancer — should also give the immortalists pause.

In the early 1990s, the identification of specific genes underlying inherited disease — such as Huntington's chorea, early-onset breast cancer, and immune deficiency syndrome — inspired researchers to devise therapies to correct the genetic malformations. So far, scientists have carried out more than 350 clinical trials of gene therapy, and not one has been an unqualified success. One eighteen-year-old patient died in a trial in 1999, and a promising French trial of therapy for inherited immune deficiency was suspended last year after three patients developed leukemia, leading the *Wall Street Journal* to proclaim that "the field seems cursed."

The record of cancer treatment is also dismal. Since 1971, when President Richard Nixon declared a "war on cancer," the annual budget for the National Cancer Institute has increased from $250 million to $5 billion. Scientists have gained a much better understanding of the molecular and genetic underpinnings of cancer, but a cure looks as remote as ever. Cancer epidemiologist John Bailar of the University of Chicago points out that overall cancer mortality rates in the United States actually rose from 1971 until the early 1990s before declining slightly over the last decade, predominantly because of a decrease in the number of male smokers. No wonder then that experts like British gerontologist Tom Kirkwood call predictions of human immortality "nonsense."

Argument: Understanding the mind is still a huge, looming challenge. When I met the British biologist Lewis Wolpert in London in 1997, he declared that *The End of Science* was "absolutely appalling!" He was particularly upset by my critique of neuroscience. The field was just beginning, not ending, he insisted. He stalked away before I could tell him that I thought his objection was fair. I actually agree that neuroscience in many ways represents science's most dynamic frontier.

Over the past decade, membership in the Society for Neuroscience has surged by almost 50 percent, to 37,500. Researchers are probing the brain with increasingly powerful tools, including superfast magnetic resonance imagers and microelectrodes that can detect the murmurs of individual brain cells. Nevertheless, the flood of data stemming from this research has failed so far to yield truly effective therapies for schizophrenia, depression, and other disorders, or a truly persuasive explanation of how brains make minds. "We're still in the tinkering stage, preparadigm and pretheoretical," says V. S. Ramachandran of the University of California at San Diego. "We're still at the same stage physics was in the nineteenth century."

The postmodern perspective applies all too well to fields that attempt to explain us to ourselves. Theories of the mind never really die; they just go in and out of fashion. One astonishingly persistent theory is psychoanalysis, which Sigmund Freud invented a century ago. "Freud . . . captivates us even now," *Newsweek* proclaimed just last March. Freud's ideas have persisted not because they have been scientifically confirmed but because a century's worth of research has not produced a paradigm powerful enough to render psychoanalysis obsolete once and for all. Freudians cannot point to unambiguous evidence of their paradigm's superiority, but neither can proponents of more modern paradigms, whether behaviorism, evolutionary psychology, or psychopharmacology.

Science's best hope for understanding the mind is to crack the neural code. Analogous to the software of a computer, the neural code is the set of rules or syntax that transforms the electrical pulses emitted by brain cells into perceptions, memories, and decisions. The neural code could yield insights into such ancient philosophical conundrums as the mind-body problem and the riddle of free will. In principle, a solution to the neural code could give us enormous power over our psyches, because we could monitor and manipulate brain cells with exquisite precision by speaking to them in their own language.

But Christof Koch of Caltech warns that the neural code may never be totally deciphered. After all, each brain is unique, and each brain keeps changing in response to new experiences, forming new synaptic connections between neurons and even — contrary to received wisdom a decade ago — growing new neurons. This mutability (or "plasticity," to use neuroscientists' preferred term) immensely complicates the search for a unified theory of the brain and mind. "It is very unlikely that the neural code will be anything as simple and as universal as the genetic code," Koch says.

Argument: If you really believe science is over, why do you still write about it? Despite my ostensible pessimism, I keep writing about science, I also teach at a science-oriented school, and I often encourage young people to become scientists. Why? First of all, I could simply be wrong — there, I've said it — that science will never again yield revelations as monumental as evolution or quantum mechanics. A team of neuroscientists may find an elegant solution to the neural code, or physicists may find a way to confirm the existence of extra dimensions.

In the realm of applied science, we may defeat aging with genetic engineering, boost our IQs with brain implants, or find a way to bypass Einstein's ban on faster-than-light travel. Although I doubt these goals are attainable, I would hate for my end-of-science prophecy to become self-fulfilling by discouraging further research. "Is there a limit to what we can ever know?" asks David Lee. "I think that this is a valid question that can only be answered by vigorously attempting to push back the frontiers."

Even if science does not achieve such monumental breakthroughs, it still offers young researchers many meaningful opportunities. In the realm of pure research, we are steadily gaining a better understanding of how galaxies form; how a single fertilized egg turns into a fruit fly or a congressman; how synaptic growth supports long-term memory. Researchers will surely also find better treatments for cancer, schizophrenia, AIDS, malaria, and other diseases; more effective methods of agricultural production; more benign sources of energy; more convenient contraception methods.

Most exciting to me, scientists might help find a solution to our most pressing problem, warfare. Many people today view warfare and militarism as inevitable outgrowths of human nature. My hope is that scientists will reject that fatalism and help us see warfare as a complex but solvable problem, like AIDS or global warming. War research — perhaps it should be called peace research — would seek ways to avoid conflict. The long-term goal would be to explore how humanity can make the transition toward permanent disarmament: the elimination of armies and the weapons they use. What could be a grander goal?

In the last century, scientists split the atom, cracked the genetic code, landed spacecraft on the moon and Mars. I have faith — yes, that word again — that scientists could help solve the problem of war. The only question is how, and how soon. Now that would be an ending worth celebrating.

WILLIAM LANGEWIESCHE

How to Get a Nuclear Bomb

FROM *The Atlantic Monthly*

HIROSHIMA WAS DESTROYED in a flash by a bomb dropped from a propeller-driven B-29 of the U.S. Army Air Force, on the warm morning of Monday, August 6, 1945. The bomb was not chemical, as bombs until then had been, but rather atomic, designed to release the energies Einstein described. It was a simple cannon-type device of the sort that today any number of people could build in a garage. It fell nose-down for forty-three seconds, and for maximum effect never hit the ground. One thousand nine hundred feet above the city, the bomb fired a lump of highly enriched uranium down a steel tube into a receiving lump of the same refined material, creating a combined uranium mass of 133 pounds. In relation to its surface area, that mass was more than enough to achieve "criticality" and allow for an uncontrollable chain of fission reactions, during which neutrons collided with uranium nuclei, releasing further neutrons in a blossoming process of self-destruction. The reactions could be sustained for just a millisecond, and they fully exploited less than two pounds of the uranium before the resulting heat forced a halt to the process through expansion. Uranium is the heaviest element on earth, almost twice as heavy as lead, and two pounds of it amounts to only about three tablespoonfuls. Nonetheless, the explosion over Hiroshima yielded a force equivalent to 15,000 tons (15 kilotons) of TNT, achieved temperatures higher than the sun's, and emitted light-speed pulses of dangerous radiation. More than 150,000 people died.

Three days later, the city of Nagasaki was hit by an even more powerful device — a sophisticated implosion-type bomb built around a softball-sized sphere of plutonium, which crossed the mass-to-surface-area threshold of criticality when it was symmetrically compressed by carefully arrayed explosives. A 22-kiloton blast resulted. Though much of the city was shielded by hills, about 70,000 people died. Quibblers claim that a demonstration offshore, or even above Tokyo harbor, might have induced the Japanese to surrender — and if not, there was another bomb at the ready. But the idea was to terrorize a nation to the maximum extent, and there is nothing like nuking civilians to achieve that effect.

The physicists who had developed these devices understood the potential for miniaturization and a simultaneous escalation in warhead yields, past the 22 kilotons of Nagasaki and indeed past 1,000 kilotons, into the multimegaton range — the realm, when multiplied, of global suicide. Moreover, they realized that the science involved, however mysterious it seemed to outsiders, had already devolved into mere problems of engineering, the knowledge of which could not be contained. Within a few years humanity would face an objective risk of annihilation — a reality that compelled those who understood it best to go public with the facts. In the months following Japan's surrender, a group of the men responsible for building the bomb — including Albert Einstein, Robert Oppenheimer, Niels Bohr, and Leo Szilard — created the Federation of American (Atomic) Scientists (FAS), to disseminate nuclear-weapons information. Washington at the time harbored the illusion that America possessed a great secret and could keep the bomb for itself to drop or not on others. The founders of FAS disagreed. The current vice president of its Strategic Security Project, an affable scientist named Ivan Oelrich, recently said to me, "The biggest secret about the atomic bomb was whether it would work or not." But after the United States exploded one, there was no longer any question in the minds of other countries. "They knew that if they did X, Y, and Z, they would have success. So in 1945 the scientists who founded this organization said, 'Look, there *is* no secret. Any physicist anywhere can figure out what we did and reproduce it. There is no secret, and there is also no defense.'"

Some of the solutions they proposed may seem quaint. Albert Einstein, for instance, called for the creation of an enlightened world government, complete with the integration of formerly hostile military staffs and the voluntary dismantling of sovereign states.

But the founders of FAS were not naive so much as desperate and brave. They said, in essence, If you knew what we know about these devices, you would agree that at any price, the practice of war must end. It was a rare call for radical change by men at the top of their game. But history shows that the future is impossible to predict. There was no exception here. After sixty years there has been no apocalypse, and a nuclear peace has so far endured for all the wrong reasons — an unenlightened standoff between the nuclear powers, each of them restrained from shooting first not by moral qualms, but by the certainty of a devastating response. Moreover, the very lack of defense that worried the scientists in 1945 turned out to *be* the defense, though treacherous because it required tit-for-tat escalations. But these are latter-day corrections to the concerns of enormously competent men, and their message is equally valid today. Detailed knowledge of nuclear-bomb making has escaped into the public domain, and the use of even a single fission device could pose an existential threat to the West.

Last winter in Moscow I spoke to an experienced cold war hand, who had skated through the collapse of the Soviet Union and now occupied a high position in the nuclear bureaucracy of the increasingly assertive Russian state. In his corduroy suit, with his bushy eyebrows and heavy, sometimes glowering face, he looked like an apparatchik from Central Casting — and he acted like one too. It was refreshing. He kept poking his finger at me and accusing Americans of losing perspective over a nuclear Iran. He wanted to do nuclear business with Iran, in electric-power generation. He wanted to do nuclear business with all sorts of countries. He claimed that with one Russian submarine reactor (fueled by high-octane uranium) he could light up all sorts of cities. He meant with electricity. He proposed a scheme to mount such reactors on barges, to be pushed to places like Indonesia and then pulled away whenever the natives ran amok. This way, he said, he could keep his uranium from fueling native bombs. He did not deny the incentives for lesser nations to acquire nuclear devices, but he thought he could handle them, or perhaps he didn't care. He said, "The Nuclear Non-Proliferation Treaty was the child of Russia and the United States. And this child was raised to fight against other countries, to resist the threat of proliferation. We're talking about the 1970s. No one thought that proliferation could come from Arab countries, from Africa, from South America. The treaty was aimed at Western Germany, at Japan. It was aimed at dissuading the developed countries from acquiring nuclear weapons — and it worked because they accepted the U.S. and Soviet nuclear umbrellas."

He was bullying history, but only by a bit. The Nuclear Non-Proliferation Treaty, or NPT, was an effort to preserve the exclusivity of a weapons club whose membership consisted originally of only five: Britain, China, France, the Soviet Union, and the United States. To other countries the treaty promised assistance with nuclear research and power generation in return for commitments to abstain from nuclear arms. It cannot be said to have "worked," as my Russian friend claimed, but it did help to slow things down. More important, and completely independent of the NPT, were the cold war alliances that, by offering retaliatory guarantees, eliminated the need for independent nuclear defense capabilities in those nations willing (or forced) to choose sides. But neither the cold war alliances nor the NPT could counter the natural appeal of these devices — their fast-track, nation-equalizing, don't-tread-on-me, flat-out awesome destructive power.

In 1946, Robert Oppenheimer sketched the problem clearly. In an essay titled "The New Weapon," he wrote: "Atomic explosives vastly increase the power of destruction per dollar spent, per man-hour invested; they profoundly upset the precarious balance between the effort necessary to destroy and the extent of the destruction." Elaborating, he wrote,

> None of these uncertainties can becloud the fact that it will cost enormously less to destroy a square mile with atomic weapons than with any weapons hitherto known to warfare. My own estimate is that the advent of such weapons will reduce the cost, certainly by more than a factor of ten, more probably by a factor of a hundred. In this respect only biological warfare would seem to offer competition for the evil that a dollar can do.

From his perch in Moscow, my Russian friend had observed the effect of Oppenheimer's truths. Continuing with his story of the NPT, he said, "Even as the U.S. and Russia offered our nuclear umbrellas, everyone understood that the weapons could never be used, because of retaliation. For us they were not wealth — they were a burden. At the same time, nuclear technology was becoming even cheaper, more efficient, and it became available to many countries. It became a useful tool especially for weak countries to satisfy their ambitions without much expense. There are no technical barriers, and no barriers to the flow of information, that can prevent it. Once a country has made the decision to become a nuclear-weapons power, it will become one regardless of any guarantees. You needn't be rich. You needn't be technically developed. You can be Pakistan, Libya, North Korea, Iran. You can be . . ." He searched for a country even more absurd in his estimation. He said, "You can be Hungary." Then he said, "At some point this change occurred. The great powers were stuck with arsenals they could not use. And nuclear weapons became the weapons of the poor."

It was a simplified view, but not entirely wrong. Certainly the argument can be made that only underdeveloped nations can now afford to use these weapons, not merely because the lives of their citizens seem to be expendable, but also because of the limitations of their nuclear arsenals, which mean that their warheads will incinerate just a few enemy cities, more or less locally, and will not likely frighten Russia and the United States into swapping strikes. The fate of the world is not at stake. Pakistan and India came close to a regional nuclear exchange in 2002, with little risk of igniting a global conflict. The core of that story, however, is that each antagonist had its own cities to protect — particularly its own capital — and each therefore had good reasons for backing down. This was a demonstration rather than a proof, but of special interest because it involved a country as backward as Pakistan. It seems to indicate as a general rule that even a stunted state is deterred by the threat of retaliation, because so long as its leaders have a government, an infrastructure, and indeed a delineated nation (not to mention their individual lives of luxury), they provide rich targets to be smashed and burned in answer to any first strike. At this point it appears that simple calculations of self-preservation should keep fingers off the triggers even in Pyongyang and Tehran. So we should be safe, relatively — but perhaps we are not.

The danger comes from a direction unforeseen in 1945, that this technology might now pass into the hands of the new stateless guerrillas, the jihadists, who offer none of the targets that have underlain our nuclear peace — no permanent infrastructure, no capital city, no country called home. The nuclear threat posed by the jihadists first surfaced in the chaos of post-Soviet Russia in the 1990s, and took full form after the fall of the World Trade Center. With so little to fear of nuclear retaliation, and having already panicked the United States into historic policy blunders, these are the rare people in a position to act.

If you were a terrorist and a bomb was your goal, how would you go about getting one? You could not bet on acquiring an existing weapon. These are held as critical national assets in fortified facilities guarded by elite troops, and they would be extremely difficult to get at, or to buy. Reports have suggested the contrary, particularly because of rumors about the penetration of organized crime into the Russian nuclear forces and about portable satchel nukes, or "suitcase bombs," which are said to have been built for the KGB in the late 1970s and 1980s and then lost into the black market following the Soviet breakup. However, the existence of suitcase bombs has never been proved, and there has never been a single verified case, anywhere, of the theft of any sort of nuclear weapon. Thefts may nonetheless have occurred, but nuclear weapons require regular maintenance, and any still lingering on the market would likely have become duds. Conversely, because these time limitations are well known, the very lack of a terrorist nuclear strike thus far tends to indicate that nothing useful was ever stolen. Either way, even if the seller could provide a functioning device, nuclear weapons in Russia and other advanced states have sophisticated electronic locks that would defeat almost any attempt to trigger them. Of course you could look to countries where less rigorous safeguards are in place, but no government handles its nuclear arsenal loosely, or would dare to create the impression that it is using surrogates to fight its nuclear wars. Even the military leaders of Pakistan, who have repeatedly demonstrated their willingness to sell this technology, would balk at allowing a constructed device to escape — if only because of the certainty that this time they would be held to account. The same concerns would almost certainly restrain North Korea.

All this should give you pause long enough to take bearings. You would do well to distinguish between your needs and those of conventional proliferators. Fledgling nuclear-weapons states have little use for just one or two bombs. To assume a convincing posture of counterstrike and deterrence, or simply to exhibit nuclear muscle, they require a significant arsenal that can be renewed and improved and grown across time. This in turn requires that they build large-scale industrial facilities to produce warhead fuel, which cannot be purchased on the international black market in sufficient quantity to sustain a nuclear-weapons production line. Manufacturing high-quality fuel is the most difficult part of any nuclear program; the NPT is meant to interfere primarily at this stage. The construction of everything else is relatively easy.

You could hardly expect to set up the facilities to manufacture nuclear fuel. (Nor could you expect any state, whether Pakistan or North Korea, to risk helping you here, either.) But you would have no need to do so. There is plenty of weapons-grade fissile material in the world today, and more is being produced all the time. Surely you could steal or buy the quantity necessary for a single garage-made device.

You would now need to decide what kind of fuel to pursue. There are really only two choices — plutonium or highly enriched uranium. Plutonium is a manmade element produced by uranium reactors. There are several forms of it, including one purpose-made for bombs. Armies favor plutonium because it can be made to go critical in very small quantities, thereby lending itself to the miniaturization of weapons. Miniaturization has obvious attractions, but it requires a level of engineering sophistication that lies beyond the capabilities of a small terrorist team. And miniaturization is not that important for your purposes. You can operate well enough with a car-sized device locked into a shipping container or loaded into a private airplane behind a couple of dedicated pilots. Furthermore, plutonium has some negatives for an operation like yours. It is not suited for use in a basic cannon-type bomb and demands instead the explosive symmetry of a Nagasaki-style implosion device. Building an implosion device would introduce complexities you would be better off avoiding, particularly without a place and the time to test the design. And plutonium is difficult to handle — sufficiently radioactive to require shielding, awkward to transport without setting off radiation detectors, and extremely dangerous even in minute quantities if it is breathed in, swallowed, or absorbed through a cut or open wound. There are plenty of people who would willingly die for the chance to nuke the United States, but within the limited pool of technicians who might join your effort, it would be impractical to expect so much. Plutonium might work as the pollutant spread by a dirty bomb, but for your project, plutonium is out.

The alternative is highly enriched uranium, or HEU, the variant of natural uranium that has been refined to contain artificially dense concentrations of the fissionable isotope U-235. Operationally it is wonderful material — the perfect fuel for a garage-made bomb. During processing, uranium takes the form of an invisible gas, a liquid, a powder, and finally a dull gray metal. It has approximately the toxicity of lead and would sicken shop workers who happened to swallow traces of it or breathe in its dust, but otherwise it is not immediately dangerous, and indeed is so mildly radioactive that it can be picked up with bare hands and, when lightly shielded, taken past many radiation monitors without setting off alarms. As one physicist in Washington suggested to me, in small masses HEU is so benign that you could sleep with it under your pillow. He warned me, however, that you could not just pile it up in your bed, or anywhere else, because the atoms of U-235 occasionally split apart spontaneously, and in doing so fire off neutrons, which within a sufficient mass of material could split enough other atoms to cause a chain reaction. Such a reaction would not amount to a military-style nuclear explosion, but it could certainly take out a few city blocks.

I asked the physicist if he wasn't concerned about giving information to terrorists. He summoned his patience visibly and said, to paraphrase it, This is Boy Scout Nuclear Merit Badge stuff. We continued our discussion. He said that the critical mass of uranium is inversely proportional to the level of enrichment. At the low end of HEU, which is considered to be 20 percent enrichment, nearly a ton would have to be combined before a stockpile could spontaneously ignite. At the high end, which is the "weapons grade" enrichment of 90 percent or greater, about 100 pounds could do the trick. I mentioned that at whatever level of enrichment, the HEU that a terrorist could acquire would by definition be made of units each consisting of less than a critical mass. I asked the physicist to imagine that a terrorist had acquired two bricks of weapons-grade HEU, each weighing 50 pounds: how far apart would he have to keep them? He said that a yard would be enough.

I had arrived in Washington from remote mountains along the Turkish border with Iran, where every night hundreds of pack-horses are led across the line by Kurdish smugglers, carrying cheap fuel for Turkish cars and opium for the European heroin market. This is the Silk Road revived, and it is one of the prime potential routes for the movement of stolen uranium. With this in mind, I told the physicist I assumed from his measure that the two bricks could be slung on either side of a saddle.

He said, "One on each side should be all right . . ." He hesitated. "But what is the moderating effect of a horse?"

I had no idea. He said, "Look, if someone's smart enough to have snuck in and gotten ahold of these two ingots of metal, he'll be smart enough to negotiate for a second horse."

But you'd probably rather not have to sneak into anywhere, or negotiate for transport, or spend cold nights squatting with peasants and dodging border patrols. Every move in this venture, every elaboration, increases the chance for something to go wrong. Furthermore, to judge from the reports that have been written about a global black market in fissile materials, perhaps you could sit on the periphery — say in Istanbul — and with relatively little risk allow the uranium to come to you.

It is difficult to get a clear picture here. Turkey is the world's grand bazaar, and given its geographic position overlooking the Middle East, it is hardly surprising that people have gone looking to sell nuclear goods there. A University of Salzburg database (formerly run by Stanford) that purports to track smuggling activity globally since 1993 lists at least twenty incidents in the vicinity of Istanbul alone. But by including intercepts of all sorts of nuclear materials, that database (like most other treatments in this business) overstates the market's ability to answer a bomb-maker's needs. In fact, the marketplace — whether in Istanbul or anywhere else — seems never to have produced what you would require. The closest instance I found in the record dates from 1998, when agents from the Russian Federal Security Bureau (the former KGB, now called the FSB) arrested nuclear workers who were plotting to steal 40 pounds of HEU. The enrichment level was never made public — an omission hinting that the uranium may well have been weapons-grade. But even then it was less than half of what you would need.

Of course, we don't know what we don't know, as we are repetitively reminded. However, the other intercepts have been minor affairs of people caught filching or hawking scraps, often of material that doesn't pass even the 20 percent mark. For a serious bomb-builder, the reports of "loose nukes" would come to sound like so much background chatter. You will have to obtain the fissile material at the source. If you look through the literature, you'll soon realize that one of the challenges is the very extent of choices. It turns out that the world is rich with fresh, safe, user-friendly HEU — a global accumulation of over 1,000 metric tons (outside of our collective 30,000 nuclear warheads) that is dispersed among hundreds of sites and separated into nicely transportable, necessarily subcritical packages. The question is how to pick some up. Here again the literature can provide guidance. Although almost all of the HEU is in some manner guarded, there are many countries where it might nonetheless be acquired, and probably nowhere better than Russia.

When post-Soviet Russia came into being, in 1991, it inherited a sprawling state industry that had provided a full range of nuclear services, including medical science, power generation, and ship propulsion — as well as the world's largest nuclear-weapons arsenal and, almost coincidentally, the world's largest inventory of surplus plutonium and HEU, maybe 600 metric tons. The physical plant consisted of several dozen research, production, and storage facilities, and especially of ten fenced and guarded nuclear cities, which housed nearly a million people, yet were nominally so secret that they did not appear on maps. But within a few years the industry was obsolete, unable to adapt to the new Russian economy, and in steep decline. The buildings were in disrepair, and morale was low because people were not being paid enough or on time. Worse, the nuclear stockpiles were apparently being neglected. There were stories of guards abandoning their posts to forage for food and of sheds containing world-ending supplies of HEU protected by padlock only. The question now, some fifteen years later, is why terrorists or criminals apparently did not then take advantage. One explanation is that they were ignorant, incompetent, and distracted. Another is that the defenses were not as weak as they appeared.

In any case, the U.S. government reacted rapidly to a perception of chaos and opportunity in post-Soviet nuclear affairs and in 1993 launched an ambitious complex of "cooperative" programs with all the former Soviet states to lessen the chance that nuclear weapons might end up in the wrong hands. The programs have blossomed into the largest part of American aid to Russia, amounting so far to several billion dollars. There have been two main efforts. The first, managed by the U.S. Department of Defense, has concentrated on getting Russia to consolidate, secure, and to some degree destroy nuclear warheads, as well as some of the missiles, aircraft, and submarines that carry them. The same programs have facilitated the spectacular denuclearization of the former Soviet Union's outlying nations. But these were the maneuverings of conventional actors following the familiar logic of strike and counterstrike. By comparison the perceived vulnerability of fissile materials in the former Soviet Union has presented the United States with a wilderness of unknowns. Securing these stocks is the second main cooperative effort. The job has been given primarily to the U.S. Department of Energy, and specifically to officials there with experience managing the American nuclear-weapons infrastructure — a group now formed into a semiautonomous agency known as the National Nuclear Safety Administration, or NNSA.

The NNSA sends managers from Washington and technicians from U.S. national laboratories to supervise the local officials. Its frontline agents tend to be hands-on technical people, impatient to pour concrete and get the jobs done. Their impulses lie primarily in an area known as Material Protection, Control and Accounting, or within the NNSA, lovingly, as MPC&A. In brief, this means locking the fissile materials down. Over the years the NNSA has identified approximately 220 buildings at 52 sites in Russia that are in dire need of treatment. That's a lot, and as a result, actually there are two treatments. The first is a stopgap measure called a "rapid upgrade." It involves bricking up the warehouse windows, installing stronger locks, fixing the fences, maybe hiring some guards. The second is a long-term fix called a "comprehensive upgrade." It often involves the full range of Americanized defenses, including crash-resistant fences, bombproof buildings, remote cameras and electronic sensors, bar-coded inventory scanners, advanced locks, well-armed and well-motivated guards, and all sorts of double- and triple-safe procedures.

Such complex constructs require constant care. Agents of the

NNSA see evidence already that the Russians are not committed to maintenance and operations, and some complained privately to me that as soon as U.S. funding ends, their elegant MPC&A systems will slip into disrepair. Nonetheless, the NNSA is supposed to wrap up the program, squeeze off the funding, and turn over all the required security upgrades to the Russians by the end of 2008. Knowledgeable observers are skeptical that this schedule can be met. They say that about a third of the identified buildings have yet to be given even rapid upgrades, that these contain about half of Russia's entire fissile-material stock, and that they sit at some of the most sensitive sites in the country — areas within the closed cities, where new warheads are assembled, and where the NNSA representatives are increasingly seen as meddlers and spies.

The agency's administrator, a portly former submarine captain and strategic-weapons negotiator named Linton F. Brooks, put it plainly to me. He said, "We are about giving governments the tools to work in those areas where governments have control." Fine. Brooks is an impressive man, and all the more so for his lack of theatrics. He did not pretend to be winning a war, or even to be fighting one, but more simply to be driving up the costs and complications for would-be nuclear bombers. The NNSA's job is to shift the odds and increase the likelihood that its opponents will fail. It cannot dictate to the Russians. It cannot operate with anything like the flexibility of the guerrillas. But of all the U.S. agencies recently engaged to suppress the nuclear threat, it does seem one of the few that may have contributed something real, even if it has to be called MPC&A.

The CIA appears to have added little to the effort. Presumably its people tag along on some missions, but they seem largely just to pursue conventional governmental information — estimating military capacities, or mapping the Russian bureaucracy in order to predict Russian reactions. I spoke to a former high U.S. official who said that during a decade spent securing stockpiles in Russia and receiving countless intelligence briefings, he had never once found information that would have helped him to calibrate the risks specific to a site. Who lives in the neighborhood? Who lives just outside? Who has just arrived? How the hell do any of them survive? What is meant here, tangibly, by *organized crime*? Is there other crime that counts? Who drives the flashy cars? What are the emotions of the people who do not? How much is known in the street about shipments to and from and between the plants? How much is known about what goes on inside? What do people think about the new wall and fences? What do they feel when they see an American flag? Now start all over again, and tell us about the nuclear technicians, the FSB agents, and the ordinary guards. Tell us about their lovers, their holidays, the furniture they dream of buying at IKEA. Tell us about their inner lives.

The official sighed with resignation. I suppose he felt what many believe, that if the United States is hit someday with an atomic bomb, it will in part be because of Washington's discomfort with informal realms — because of a blindness to the street, amply demonstrated in recent times, which will have allowed some bomb-builder the maneuvering room to get the job done.

If you wanted a bomb, you would need this very thing the CIA has lost — a feel for the street. I flew to Ekaterinburg, a Russian city on the Siberian side of the Urals, two time zones east of Moscow. Ekaterinburg is where the czar and his family were liquidated, where the American U-2 pilot Gary Powers was shot down, and where Boris Yeltsin got his start. It has a single-line metro, a small downtown, and a few large hotels. Within a few hours' drive, a visitor can arrive at the perimeter walls and fences of five of Russia's ten closed nuclear cities. They are primarily production sites, and between them they contain all manner of nuclear goods, including warheads in various stages of assembly and several hundred tons of excellent fissile material. So sensitive are the nuclear cities and other defense sites in the vicinity that the entire region was closed to outsiders during Soviet times. Since the Soviet Union was itself largely closed and compartmentalized, the nuclear cities stood within concentric layers of defenses like fortresses within fortresses, like nested Russian dolls. Still more pervasive defenses existed in the residents' minds. They were expected to inform on their neighbors, as they themselves expected to be informed on, and to rely on all the nested gulags to keep everyone in line.

Describing those years in heavily accented English, a Russian plant manager said to me, "All nuclear material was secret. *State* secret! Anyone stealing nuclear material in Soviet Union was committing state crime. He became state criminal! So there was fear.

Real fear. If something got lost somewhere — maybe a piece of paper, or materials, or there was mismatch in balance of plutonium — a person understood that he would be exiled forever." Hesitating over the right words, he said, "But then when this . . . *change* took place, of course people felt more . . . *freedom,* I would say."

He was using the word *freedom* for the American ear, and making a familiar argument for the extension of foreign aid. The closed cities and nuclear facilities around Ekaterinburg have received a large portion of the U.S. dollars spent on Russian security upgrades, yet they are the subject of continued wariness by the NNSA and are of still greater concern to independent critics in the United States, who insist that they remain acutely vulnerable to terrorist theft. Take, for example, the closed city of Ozersk, a community of 85,000 people, whose existence was so hushed under the Soviets that it was not allowed a proper name and was referred to only by its post office box numbers — first No. 40, then No. 65 — in Chelyabinsk, an open city 44 miles away. The nomenclature remains confused. Ozersk is often and mistakenly called Mayak, for its nuclear production area — an industrial complex within the city's perimeter, a few miles from the residential center. The Mayak Production Association employed 14,500 people as of 2001, and since 1945 has been in the business primarily of processing HEU, plutonium, and tritium for nuclear warheads. Recently it has also thrown the business into reverse, as one of two sites in Russia where fissile material from existing warheads is extracted before shipment to another closed city for blending down. Many tons of the highest-quality weapons-grade HEU and plutonium are present at the site.

And *nyet,* this is not state secret.

Divulging it is not state crime.

Start with the all-American fact that Mayak is the location of the recently completed "Plutonium Palace," a heavily fortified $350 million warehouse, which was paid for by the U.S. Congress and has therefore been heavily publicized. The facility was designed to hold as much as 40 percent of the Russian military's excess fissile material. For now the storage vaults remain empty because of technical and bureaucratic disputes, as well as, perhaps, a sense, in a Russia on the rise, of not wanting to place nuclear assets quite so far out of reach just yet. Nonetheless, there is little doubt that such a beautiful facility will eventually be used — and that the world will be better off when it is.

For those trying to counter the threat of nuclear terrorism, however, the possibility exists that the Palace is just another Maginot fort, a strong point that can be neatly ignored by the new strategists of war. The fort will neither reduce nor protect the large quantities of weapons-grade materials elsewhere in Ozersk. NNSA technicians have struggled to fill the gaps, installing some cameras and radiation monitors and strengthening some floors and walls — but only in a few of the many buildings where they believe such work is needed, and only intermittently, as the Russians have allowed.

By perverse logic, therefore, Ozersk comes highly recommended to anyone in pursuit of a bomb. The facilities are tucked away two hours south of Ekaterinburg, down unmarked back roads, on a forested plateau of lakes and small rivers that would seem idyllic were it not for a number of decimated industrial towns and villages. Ozersk's perimeter is large. It encompasses more than 50 square miles and includes the city itself, the Mayak facilities, a network of paved and dirt roads, an internal bus service, one large internal checkpoint, multiple railroad lines, burial sites for radioactive waste, multiple radioactive lakes, some radioactive swamps, and a lot of radioactive forest. The main gate is on the north side. It has pens where Interior Ministry troops check vehicles going in or out and verify people's papers. The perimeter is marked by twin parallel chain-link barbed-wire fences, separated by a chemically defoliated strip that is apparently not mined or raked or checked for tracks even in the snow. The fences are in reasonable repair but in many places have no road beside them, and they show no signs of being patrolled. If they are monitored remotely, it is safe to assume that they are not monitored well. On the south side of the site, they run through miles of empty forest. You could cross them almost anywhere with little immediate risk, though odds are you would eventually be caught, and the consequences would be severe. And to what advantage? You wouldn't learn much by walking the streets. You might be able to send agents who could establish residency, but that would require too much time and offer a poor chance of success. Better to back off and think things through, particularly since the essentials can be known from the outside.

Ozersk is in the nuclear business and nothing else. Its weapons-usable HEU is kept in Mayak as oxidized powder, flat metal pucks, elongated ingots, and finely machined warhead hemispheres. Each form is stored in a different type of steel container. The containers are light, because the shielding they contain is minimal. They are sealed but not locked. They sit in vaults or ordinary storage rooms. In addition to the standard containers, there are larger, brightly colored shipping containers, used to transport the material to and from other nuclear cities and sites, by truck and rail. Empty shipping containers are sometimes kept outdoors. High-resolution satellite photographs freely available on the Internet show them stacked in the yards and in other ways help to identify the buildings in the complex. It does not matter that the photographs are old. Any of thousands of ordinary workers at Mayak could update them and also provide information on material-processing schedules, NNSA upgrades, broken cameras, the patterns of the night shifts, and the locations of guards who use narcotics or drink on the job. It would not be difficult to find such an informant — for instance, in the taverns of Chelyabinsk. Afterward the action would have to be fast and accurate. Moving through the forest from the east or south perimeter fences, a raiding party on foot could hit any of the buildings within two hours. This would hardly be a sure thing. Ten years ago would have been a better time to try. But with luck it could still be done.

Or knowledgeable Americans urgently believe so. Some of the best of them work for a richly endowed organization known as the Nuclear Threat Initiative (NTI), which was founded by former senator Sam Nunn and the CNN mogul Ted Turner. The director of NTI's efforts in the former Soviet states is a smart, no-nonsense Washington insider named Laura Holgate, who spent years as a senior U.S. official grappling with Russian nuclear-weapons security. When we met, at NTI's buttoned-up headquarters in Washington, she repeated a colleague's anecdote about a certain closed city where informal parking lots have sprung up outside of holes in the fence, because workers prefer not to bother with the gate. But I wanted to know about Ozersk, which seems to be more orderly, perhaps because of the quantities of bomb materials there. I asked about the security upgrades. She answered indirectly. "It's a very challenging environment to work in. You set up a radiation detector, and it's constantly going off, mostly false alarms. There's so much clandestine movement of other stuff — cooking oil, insulation, you name it. So how are you supposed to catch the illicit movement of material as unradioactive as HEU? They put a little box of uranium in the corner, and they're fine." Moreover, the Russian government, she implied, can be somewhat dismissive of such matters.

Later, in Russia, I spoke to a modest American technician with a decade of experience in the secret cities of the Urals. He knew enough of Washington to want to remain anonymous but did not seem interested in diplomatic maneuvering or nuclear policy. He was a detail man, a practitioner. He said, "Some of what the NNSA puts in is worthwhile, maybe. And other stuff seems, ah, not to get used. I took a tour one time, and they had installed a radiation monitor, a portal, for traffic going in and out of the site. So I say, 'So, how's the portal working?'

"And this Russian says, 'Oh, we shut it off most of the time.'

"'Why?'

"'Because it's always going off.'

"'How come it always goes off?'

"He says, 'Well . . . it's the people on the buses. People go fishing in the lake, and when they catch fish and bring them out on the bus, they set off the radiation monitor. And then we've got to respond.'"

The technician laughed at his own story. Just when you think you've nabbed a terrorist, what you've really nabbed is a radioactive fish. After ten years around Russia's nuclear facilities, he was not concerned. He said, "But you know, things like radiation portals? That's just *our* method of security. They've got their own methods, which are probably as effective."

"You mean human intelligence?"

"Yeah. It's more *people*-intensive, the way they do it."

Others have described Ozersk's human side in dire terms — children selling drugs in the schools, mobsters in the construction and trucking business, large numbers of unvetted Central Asians arriving to do menial work that before would have been reserved for loyal Slavs. The worst of the lot are the soldiers whose duty is to guard the site. Called "the dregs of the dregs" by some critics, they are second-round conscripts, picked up by the Interior Ministry only after they have been rejected by the army. They in no sense constitute an elite corps, as the ministry sometimes claims. NTI has catalogued a string of known incidents of Mayak guards killing each other, committing suicide, stealing weapons, running away, buying narcotics, drinking on duty, and in one case imbibing a bottle of antifreeze and dying.

The problem for you, in your quest for a bomb, is that even these soldiers will fight. This is less a possibility than a fact. They will fight whether sober or drunk. Their presence at Ozersk means that no raiding party will be able to hit without provoking a noisy response. Still, American specialists have warned that a large and coordinated attack could overwhelm or counter traditional Russian defenses, pointing to the Chechen seizures of a Moscow theater in 2002 and a Beslan school in 2004 as evidence that terrorists within Russia already possess the wherewithal.

If you wanted a bomb, however, you would have to be concerned not only with getting your hands on fissile material but also with getting away. The getaway factor is not often raised in the United States, where the prevailing view is of a Wild West environment in which a nuclear bandit would need only to ride out of town in order to vanish safely. But up close to Ozersk, I kept running across evidence of a powerful and self-confident Russia that seemed more like its old autocratic self, and at least selectively under control.

The nearest international border to Ozersk is with Kazakhstan, only four hours' drive to the south, but the route is poor, because it runs through Chelyabinsk and crosses border checkpoints using roads that can easily be blocked. A safer plan would be to head southwest, 1,200 miles or more, for the Caspian Sea or the Caucasus, with the goal of smuggling the uranium to Turkey through Azerbaijan, Armenia, Georgia, or northern Iran. The trip, however, would require at least three days just for the escape from Russia — an interval that becomes the minimum head start required before the loss of material is noticed at Mayak. And raiders would be lucky to have three hours. In good conscience, therefore (and perhaps with some relief), a rational bomb-maker would abandon any idea of commando heroics.

But that is not to say you should give up on Ozersk. Possibilities abound for an inside job. A culture of mobsterism and corruption broadens the opportunities for recruiting thieves. Insiders could neutralize any practical defenses, pass undetected through the gates with a load of unshielded HEU, and provide a getaway team with a head start that could be measured in weeks or months — perhaps right up to the time an assembled bomb ignites. Linton Brooks described the possibility of insider theft as the greatest challenge facing the NNSA in Russia today. The solutions, which at best can only be partial, consist of efforts to complicate the task of would-be thieves, requiring them to bring more people into a conspiracy and to operate with larger teams.

An agent with deep experience in these matters said, "You try to make it more difficult by putting in doors that require two people to open. You put in video surveillance. You put in watchers to watch the watchers. You put in accounting systems to bring the facilities from paper ledgers into the bar-code era. But there are very deep-seated cultural issues here. In the Soviet days you were a very trusted person, an elite person, if you were working in these facilities. And now we come along and say, 'Okay, you're not allowed to go into the vault by yourself anymore.' That can take a while to sink in. Also, there's a different perception of rules and regulations. The rule of law is not looked at the same way. There's more skepticism in Russia. It's a very complicated problem. Bricking up windows is part of the solution, but it's not everything."

It's not as if the Russians don't already know how to build strong rooms. I once mentioned to the American technician in the Urals that even ordinary doors there are heavy. He said, "One of the guys at the plant said to me, 'Jack, I watch American movie last night.'" The technician imitated a Russian accent. "'And I see something strange. Drug police. They kick in door! In Russia, this not possible!'"

Or maybe in Russia, this not necessary.

With 100 pounds of stolen HEU split between a couple of knapsacks and a healthy head start on Russian security forces, you would not have to worry much about getting caught by Americans. The United States claims to be building a layered defense, but the only layer that amounts to much is the NNSA's securing of stockpiles — and you have just penetrated it with the help of workers on the site. At this point the American defenses fall spectacularly apart. The reasons are ultimately institutional and complex, but initially may be as simple as confusion created by the expanding geometry of choices for anyone carrying stolen HEU toward an assembly point for a bomb. Will you go left or right? The roads keep forking, and you will often turn one way or the other for no more discernible reason than the fact that forward motion requires the choice. American officials who would try to stop you face an infinite braid that becomes the measure of a hostile, corrupt, and anarchic world.

In Washington I spoke to one of the many officials in town who, though carrying out their assignments perfectly well, are too nervous about domestic politics to risk being identified. He described an NNSA effort called the Second Line of Defense, which installs radiation monitors at border-crossing points throughout the former Soviet Union. The program is most fully developed in Georgia, a skeletal nation threatened by separatist enclaves and barely able to keep warm through the winters. The official said, "The U.S. Corps of Engineers is working with Customs to build whole new border-crossing facilities in Georgia, and we're in the process of installing upgraded radiation monitors, in conjunction with that."

I said, "I guess I can see the logic of the radiation monitors, but why are we building customs stations for the Georgians?"

"It's a joint venture to try to control smuggling."

Georgia is one of the most corrupt nations on earth. Many of its politicians are crooks. Its officials routinely steal. Its economy is based almost entirely on black markets. Its people have nothing to survive on if they do not hustle. I said, "Why should we care about ordinary smuggling in Georgia? Cigarettes? Vodka? Fuel? For that matter, narcotics?" What I meant was, if even the ordinary black markets of Georgia are seen as a threat, where does the impulse to impose order end?

He stayed in his lane but executed a tidy reversal. He said, "It's a good question. And I don't know that I have the answer, but the genesis anyway was nuclear. So I guess in conjunction with our new equipment, it makes sense to upgrade the stations."

Hoping to get some notion of the realities, I went to a model Customs project, a station the Americans call "Red Bridge," which stands at the main crossing point between Georgia and Azerbaijan. The first improvement there was the construction of a housing compound where for several years before, border guards had been living in tents and cooking over open fires. By the time the U.S. Department of Homeland Security was finished with it, in 2003, the compound consisted of five single-family houses, a barracks, a dining hall, an administration building, a vehicle-maintenance garage, a warehouse for supplies, an armory, various utility buildings, a dog kennel, two water-storage buildings, a sewage plant, an electrical substation, perimeter fences, two guard towers, two helipads, a sports field, a separate soccer field, some new paved roads and parking lots, and, of course, a parade ground. After the official opening, a Customs and Border Protection newsletter asked the CBP program manager, James Kelly, "Is it as grand as it sounds?"

"Indeed it is," Kelly said. "It was built to Western standards."

Golly, and that was just phase one. The United States then shifted its spending down the road to the port of entry itself. Over the following two years, the CBP oversaw the construction of a $2.2 million facility, whose primary purpose would be to "help Georgia become a more effective partner in the worldwide effort to control the passage of terrorists and their weapons." This time the improvements included a six-lane roadway, inspector booths, cargo-inspection areas, closed-circuit remote-control television cameras, lots of computers, a high-frequency long-distance radio-communication system, and the crowning glory — a beautiful air-conditioned two-story stucco building with spaces for processing the grateful public, as well as detention cells, back offices, a dormitory wing for Georgian customs agents, private sleeping rooms for American officials and other VIPs, and a second-floor patio, apparently for sitting outside on warm evenings.

The day I got there was too chilly for that. A short line of trucks crept in from Azerbaijan, and a few cars passed through in the opposite direction. The new radiation monitors were not yet in place, but within a few weeks they were certain to be. I asked to see the control room, which is meant to be the nerve center of the operation. There was a delay until the key could be found, whereupon the chief showed me in. The room was a windowless refuge, empty except for a few chairs, some computers, and a bank of flat-screen displays. Someone had left a video game on. Someone had left a magazine. Empty cardboard boxes were stacked in a corner. After a while an eager young man came in and demonstrated how a camera on the roof could be swiveled and zoomed.

The anonymous official in Washington tried to explain why such improvements matter. I had asked him why anyone carrying a bomb's worth of HEU would choose to go through any official gate anywhere. He did not predict that anyone would. Using Georgia as an example, he said, "The good news is that there aren't very many of these border crossings. And to the extent we've got them covered, we force people to use horses, I guess." Horses, mules, donkeys, tractors, dirt bikes, plucky little Ladas — whatever. A bit of off-road travel wouldn't seem like much of an inconvenience to nuclear terrorists on the move, and American officials know it well. The problem is that U.S. agencies, when pressured by conflicting mandates and forced to work with corrupt and dysfunctional local governments, essentially throw up their hands at the complexity of it all and abandon the fight in advance.

In this context, Red Bridge is not just a customs post but a premature surrender — and a typical one. Faced with the need to put systems in place that will function day after day to identify unexpected nuclear smugglers, America turns to the uniformed agents of local governments, loads them down with air-conditioned buildings and gadgetry, and then asks them to sit in a closed room watching for information from television cameras and radiation monitors. To make things worse, the failure is not individual but collective, and therefore difficult to correct. It appears to include even the clandestine services, some of which are chasing al-Qaeda around, but none of which shows signs of wanting to lay traplines through the backcountry on the off chance — highly unlikely in any given location — of snaring two knapsacks' worth of dull gray metal. Of course it is possible that they are doing this and being so discreet that for once they leave no evidence of their passage, but if I had to move a load of HEU across international borders, I would gamble that they are not.

It is as if the U.S. government, when looking at a world map for purposes of HEU interdiction, has declared entire regions to be off-limits to anything other than fictions. As noted, those backcountry regions are where some of the world's principal opium routes lie and, for independent reasons, are the areas through which stolen HEU seems most likely to pass. It is generally assumed that for the right price, opium traffickers will provide transportation, lodging, and expert advice to nuclear terrorists moving through. Indeed, the persistence of the drug trade worldwide is often used as an object lesson in the difficulty of trying to stop the smuggling of nuclear materials. A tired joke is that the best way to smuggle an atomic bomb is to put it into a bale of marijuana. For the United States the proximity between the two trades is an unfortunate coincidence, but it could be turned into a fortunate one, through quiet conversations with a few key people. If you wanted a bomb, you should of course be having quiet conversations with those same key people.

Finding them merely requires casual exploration along the preexisting lines of defense, especially along the national boundaries that cross the smuggling routes. The work would involve poking around meekly, sometimes with an amateur translator and guide. The purpose would not be to recruit peasant armies and spies, but to get a feel for the local functioning of power. In most areas there are only two or three people at the top, and they tend to be at once aggressive and benevolent men with interests larger than just the movement of drugs. Their names would quickly become apparent. Some might be dangerous to approach, but most would be hospitable to strangers. On the second or third trip back, a U.S. agent might make it known that if ever a load of HEU showed up, a large reward would be paid.

Some months ago I made the first of two brief trips through the mountains of far eastern Turkey, in the Kurdish hinterland along the border with Iran. This is the prime smuggling country I mentioned earlier. The local villages may seem sleepy, but they are wide awake. It became apparent to me that the entire region is tightly sewed up, that nothing moves here without notice, and that any sustained activity requires approval. The authority is not the Turkish government. The army is here to fight the sporadic guerrilla war against Kurdish separatists, who occasionally ambush a patrol or plant a mine, then retreat higher into the mountains. The main garrison is in a tiny town called Bascale, which is better known for its heroin labs. When A. Q. Khan was building Pakistan's nuclear-weapons facilities, some of the more sensitive parts passed through here.

Bascale has a large yellow house in a central compound, which belongs to the chief of a clan that is dominant for miles around, an extended family named Ertosi, which consists of 150,000 to 200,000 people. One afternoon I was invited in for tea. The host was a black-bearded, heavyset middle-aged man who had arrived from Ankara for a few days. He was a son of the chief and so was important too. He had guards and flunkies around. We sat against the walls in a large bare room, with photographs of dead Ertosis at one end and a big-screen television at the other. The television was tuned to a banned Kurdish station, broadcast via satellite from London. My translator introduced me as an English teacher on holiday, which threw our host a little off balance, I think. He wore a heavy gold chain and a diamond ring. He had an array of cell phones in front of him on the floor, several of which rang. Between calls, I asked him about the diesel trade. He said a little, but then clammed up when I got into details. To my translator he said, "Why does an English teacher want to know so much?"

Weeks later I returned to the area and sought out a certain subclan leader I had heard about. He was a powerful man, the chief of the 20,000 Ertosis who occupy the most active stretch of the border. I found him in a hamlet high on a mountainside. It was late at night, and the air was cold. We sat among ten of his men in a small stone room around a wood-burning stove. He was a small man, sixty years old, with sharp hooded eyes and a hooked nose. He was dressed like a Kurdish peasant, with a tweed jacket down to his knees. His hands were hardened by manual labor. From the ease with which he gave orders and the deference shown to him by his men, it was clear that he had been laying out law for many years. We avoided mention of narcotics or HEU and talked instead in a transparent code about the business in "diesel fuel." He said, "The government cannot control the border. The Kurds just naturally do." No stranger could cross without his knowledge.

Now back to the project at hand. If you wanted to build a bomb and paid attention to men like the clan leaders in the mountains, you should have no trouble getting the bomb fuel through.

Then comes the problem of assembly. You should take it as a given that no nation would dare to be associated with the construction of an independent atomic bomb. Not North Korea, not Sudan, not Iran. The certainty of retribution after its use far outweighs whatever benefit might be gained. Moreover, you could never trust those governments not to wait until the end and confiscate the goods. So the work will have to be carried out in secret —

and in some private machine shop perhaps no larger than a five-car garage. The shop will contain numerically controlled milling machines and lathes, as well as other expensive equipment, and will require a plausible explanation — a front company set up to manufacture, say, transmission components. The best location would seem to be in some Third World city, where governmental control is lax, corruption is rampant, and the noise from the shop will be masked by industrial activities nearby. Take your pick: Mombasa, Karachi, Mumbai, Jakarta, Mexico City, and a finite list of others. You might as well choose Istanbul, because it's close to where you'd be coming from.

Construction of the bomb would take maybe four months. The size of your technical team would depend on the form of HEU. At the minimum it would consist of a nuclear physicist or engineer, a couple of skilled machinists, an explosives expert, and perhaps an electronics person, for the trigger. The essentials of the work are easy to grasp. A cannon-type bomb employs two HEU hemispheres which when joined together form a single sphere. A sphere is the geometric shape that provides the smallest surface area in relation to mass, lowering the threshold of criticality. The receiving hemisphere is hollow on one side. The other hemisphere, known as the bullet or plug, has a perfectly matched convex surface on the opposing side. It sits in wait at the top of a steel barrel sometimes as long as a stepladder is high. When the time comes to detonate the bomb, a chemical propellant shoots the plug down the barrel. The ideal velocity is 1,000 meters per second, about as fast as a high-speed rifle round. The goal is to slam the masses together before the nuclear reaction has a chance to start. Once they have joined, the HEU can be trusted to do its job, though with a primitive, garage-made device there will be some uncertainty about the explosive yield. A common misperception about terrorist bomb-makers is that they would try to imitate a military device and would therefore require a level of expertise found only in government laboratories — but you would not care whether New York was hit with a 10- or a 20-kiloton blast.

Even so, the construction of a bomb is not a casual project. The machinery, the noise, and especially the likely presence of team members who are not locals provide the United States with the last chance of self-defense. A city like Istanbul, which appears from a distance to be anarchic and is famously resistant to central authority, is up close and in practice a patchwork of tightly knit communities with something of the organic power structures of the borderlands. In even the most chaotic neighborhoods, where industrial shops are mixed among illegal apartment blocks and communities of impoverished newcomers and squatters, it would be difficult to keep the neighbors from asking inconvenient questions. The same is true in Mombasa, Karachi, and every other city where a bomb could conceivably be built. A United States government that could lay traplines in their slums would have a much better chance of stopping a terrorist attack than any amount of naval maneuvering or bureaucratic restructuring can provide.

In the end, if you wanted a bomb and calculated the odds, you would have to admit that they were stacked against you, simply because of how the world works — and that this may be why others like you, if there have been any, have so far not succeeded. You would understand, though, that the odds are not impossible. You would of course have many concerns as you moved ahead. But perhaps the thing that should worry you the least is the American government's war on terror.

JONAH LEHRER

The Effeminate Sheep

FROM *Seed*

MALE BIGHORN SHEEP live in what are often called "homosexual societies." They bond through genital licking and anal intercourse, which often ends in ejaculation. If a male sheep chooses not to have gay sex, it becomes a social outcast. Ironically, scientists call such strait-laced males "effeminate."

Giraffes have all-male orgies. So do bottlenose dolphins, killer whales, gray whales, and West Indian manatees. Japanese macaques, on the other hand, are ardent lesbians; the females enthusiastically mount each other. Bonobos, one of our closest primate relatives, are similar, except that their lesbian sexual encounters occur every two hours. Male bonobos engage in "penis fencing," which leads, surprisingly enough, to ejaculation. They also give each other genital massages.

As this list of activities suggests, having homosexual sex is the biological equivalent of apple pie: everybody likes it. At last count, over 450 different vertebrate species could be beheaded in Saudi Arabia. You name it, there's a vertebrate out there that does it. Nevertheless, most biologists continue to regard homosexuality as a sexual outlier. According to evolutionary theory, being gay is little more than a maladaptive behavior.

Joan Roughgarden, a professor of biology at Stanford University, wants to change that perception. After cataloging the wealth of homosexual behavior in the animal kingdom two years ago in her controversial book *Evolution's Rainbow* — and weathering critiques that, she says, stemmed largely from her being transgendered — Roughgarden has set about replacing Darwinian sexual selection with a new explanation of sex. For too long, she says, biology has neglected evidence that mating isn't only about multiplying. Sometimes, as in the case of all those gay sheep, dolphins, and primates, animals have sex just for fun or to cement their social bonds. Homosexuality, Roughgarden says, is an essential part of biology and can no longer be dismissed. By using the queer to untangle the straight, Roughgarden's theories have the potential to usher in a scientific sexual revolution.

Darwin's theory of sex began with an observation about peacocks. For a man who liked to see the world in terms of functional adaptations, the tails of male peacocks seemed like a useless absurdity. Why would nature invest in such a baroque display of feathers? Did male peacocks *want* to be eaten by predators?

Darwin's hypothesis was typically brilliant: the peacocks did it for the sake of reproduction. The male's fancy tail entranced the staid peahen. Darwin used this idea to explain the biological quirks that natural selection couldn't explain. If a trait wasn't in the service of survival, then it was probably in the service of seduction. Furthermore, the mechanics of sex helped explain why the genders were so different. Because eggs are expensive and sperm are cheap, "Males of almost all animals have stronger passions than females," Darwin wrote. "The female . . . with the rarest of exceptions is less eager than the male . . . she is coy." Darwin is telling the familiar Mars and Venus story: men want sex while women want to cuddle. Females, by choosing whom to bed, impose sexual selection on the species.

Darwin's theory of sex has been biological dogma ever since he postulated why peacocks flirt. His gendered view of life has become a centerpiece of evolution, one of his great scientific legacies. The culture wars over evolution and common descent notwithstanding, Darwin's theory of sexual selection has been thoroughly assimilated into mass culture. From sitcoms to beer ads, our coital "instincts" are constantly reaffirmed. Females are wary, and males are horny. Sex is this simple. Or is it?

Indeed, biology now knows better. Nobody is hornier than a female macaque or bonobo (which mount the males because the males are too exhausted to continue the fornication). Peacocks are actually the exception, not the rule.

Roughgarden first began thinking Darwin may have been in error after she attended the 1997 Gay Pride parade in San Francisco, where she had gone to walk alongside a float in support of transgendered people. Although she had lived her first fifty-two years as a man, Roughgarden was about to become a woman. The decision hadn't been easy. For one thing, she was worried about losing her job as a tenured professor of biology at Stanford. (The fear turned out to be unfounded.)

After living for a year in Santa Barbara while undergoing the "physical aspects of the transition," Roughgarden returned to Stanford in the spring of 1999 and decided to write a book about the biology of sexual diversity. In particular, she wanted to answer the question that had first surfaced in her mind back in 1997. "When I was at that Gay Pride parade," Roughgarden remembers, "I was just stunned by the sheer magnitude of the LGBT [Lesbian, Gay, Bisexual, Transgender] population. Because I'm a biologist, I started asking myself some difficult questions. My discipline teaches that homosexuality is some sort of anomaly. But if the purpose of sexual contact is just reproduction, as Darwin believed, then why do all these gay people exist? A lot of biologists assume that they are somehow defective, that some developmental error or environmental influence has misdirected their sexual orientation. If so, gay and lesbian people are a mistake that should have been corrected a long time ago. But this hasn't happened. That's when I had my epiphany. When scientific theory says something's wrong with so many people, perhaps the theory is wrong, not the people."

The resulting book, *Evolution's Rainbow,* was an audacious attack on Darwin's theory of sexual selection. To make her case, Roughgarden filled the text with a staggering collection of animal perversities, from the penises of female spotted hyenas to the ménage à trois tactics of bluegill sunfish. As Roughgarden put it, "What's coming out [in the past ten to fifteen years] is to the rest of the species what the Kinsey Report was to humans."

According to Roughgarden, classic sexual selection can't account for these strange carnal habits. After all, Darwin imagined sex as a relatively straightforward transaction. Males compete for females. Evolutionary success is defined by the quantity of offspring. Thus, any distractions from the business of making babies — distractions like homosexuality, masturbation, etc. — are precious wastes of fluids. You'd think by now, several hundred million years after sex began, nature would have done away with such inefficiencies, and males and females would only act to maximize rates of sexual reproduction.

But the opposite has happened. Instead of copulation's becoming more functional and straightforward, it has only gotten weirder as species have evolved — more sodomy and other frivolous pleasures that are useless for propagating the species. The more socially complex the animal, the more sexual "deviance" it exhibits. Look at primates: compared to our closest relatives, contemporary, Westernized *Homo sapiens* are the staid ones.

Despite this new evidence, sexual selection theory is still stuck in the nineteenth century. The Victorian peacock remains the standard-bearer. But as far as Roughgarden is concerned, that's bad science: "The time has come to declare that sexual theory is indeed false and to stop shoehorning one exception after another into a sexual selection framework . . . To do otherwise suggests that sexual selection theory is unfalsifiable, not subject to refutation."

Roughgarden is an ambitious scientist. She believes it is impossible to comprehend the diversity of sexuality without disowning Darwin. Although she isn't the first biologist to condemn sexual selection — Darwin's theory has never been very popular with feminists — she is unusually vocal about cataloguing his empirical errors. "When I began, I didn't set out to criticize Darwin," she says. "But I quickly realized that most scientists are pretty dismissive about same-sex sexuality in vertebrates. They think these animals are just having fun or practicing. As long as scientists clung to this old dogma, homosexuality would always be this funny anomaly you didn't have to account for."

Roughgarden's first order of business was proving that homosexuality isn't a maladaptive trait. At first glance, this seems like a futile endeavor. Being gay clearly makes individuals less likely to pass on their genes, a major biological faux pas. From the perspective of evolution, homosexual behavior has always been a genetic dead end, something that has to be explained away.

But Roughgarden believes that biologists have it backwards. Given the pervasive presence of homosexuality throughout the animal kingdom, same-sex partnering must be an adaptive trait that's been carefully preserved by natural selection. As Roughgarden points out, "a 'common genetic disease' is a contradiction in terms, and homosexuality is three to four orders of magnitude more common than true genetic diseases such as Huntington's disease."

So how might homosexuality be good for us? Any concept of sexual selection that emphasizes the selfish propagation of genes and sperm won't be able to account for the abundance of nonheterosexual sex. All those gay penguins and persons will remain inexplicable. However, if one looks at homosexuality from the perspective of a community, one can begin to see why nature might foster a variety of sexual interactions.

According to Roughgarden, gayness is a necessary side effect of getting along. Homosexuality evolved in tandem with vertebrate societies, in which a motley group of individuals has to either live together or die alone. In fact, Roughgarden even argues that homosexuality is a defining feature of advanced animal communities, which require communal bonds in order to function. "The more complex and sophisticated a social system is," she writes, "the more likely it is to have homosexuality intermixed with heterosexuality."

Japanese macaques, an Old World primate, illustrate this principle perfectly. Macaque society revolves around females, who form intricate dominance hierarchies within a given group. Males are transient. To help maintain the necessary social networks, female macaques engage in rampant lesbianism. These friendly copulations, which can last up to four days, form the bedrock of macaque society, preventing unnecessary violence and aggression. Females that sleep together will even defend each other from the unwanted advances of male macaques. In fact, behavioral scientist Paul Vasey has found that females will choose to mate with another female, as opposed to a horny male, 92.5 percent of the time. While this lesbianism probably decreases reproductive success for macaques in the short term, in the long run it is clearly beneficial for the species, since it fosters social stability. "Same-sex sexuality is just another way of maintaining physical intimacy," Roughgarden says. "It's like grooming, except we have lots of pleasure neurons in our genitals. When animals exhibit homosexual behavior, they are just using their genitals for a socially significant purpose."

Roughgarden is now using this model of homosexuality to reimagine heterosexuality. Her conclusions, published last February in *Science,* are predictably controversial. While Darwin saw males and females as locked in conflict, acting out the ancient battle of their gametes, Roughgarden describes sexual partners as a model of solidarity. "This whole view of the sexes as being at war is just so flawed from the start. First of all, there are all these empirical exceptions, like homosexuality. And then there's the logical inconsistency of it all. Why would a male ever jettison control of his evolutionary destiny? Why would he entrust females to serendipitously raise their shared young? The fact is, males and females are committed to cooperate."

Consider the Eurasian oystercatcher, a shorebird that enjoys feasting on shellfish. A consistent minority of oystercatcher families are polygynous, in which a lucky male mates with two different females simultaneously. These threesomes come in two different flavors: aggressive and cooperative. In an aggressive threesome, the females are at war; they attack each other frequently and try to disrupt the egg-laying process of their fellow spouse. So far, so Darwinian: life is nasty, brutish, and short. However, the cooperative threesome is everything Darwin didn't expect. These females share a nest, mate with each other several times a day, and preen their feathers together. It's domestic bliss.

In Roughgarden's *Science* paper, she uses "cooperative game theory" to elucidate the diverse mating habits of the oystercatcher. Whereas Darwin held that conflict was the natural state of life (we are all Hobbesian bullies at heart), Roughgarden sees cooperation as our default position. This makes mathematical sense: the family that sleeps together has more offspring. Why, then, do oystercatcher females occasionally engage in all-out war? According to Roughgarden, violence occurs when "social negotiations" break down. Although the birds really want to get along (who doesn't like being preened?), something goes awry. The end result is risky violence, in which one female or both will end the breeding season without an egg.

The advantage of Roughgarden's new theory is that it can explain a wider spectrum of sexual behaviors than Darwinian sexual selection. Lesbian oystercatchers and gay mountain sheep? Their homosexuality is just a prelude to social cooperation, a pleasurable way of avoiding wanton conflict. But what about the peacock and all those other examples of sexual dimorphism? According to

Roughgarden, "Expensive, functionally useless badges like the peacock's tail . . . are admission tickets": they just get you in the door. If you don't have a ticket, you are ruthlessly denied breeding rights, like an uncool kid at the prom.

Of course, most humans don't see sex as a way of maintaining the social contract. Our lust doesn't seem logical, especially when that logic involves the abstruse calculations of game theory. Furthermore, it's strange for most people to think of themselves as naturally bisexual. Being gay or straight seems to be an intrinsic and implacable part of our identity. Roughgarden disagrees. "In our culture, we assume that there is a straight-gay binary, and that you are either one or the other. But if you look at vertebrates, that just isn't the case. You will almost never find animals or primates that are exclusively gay. Other human cultures show the same thing." Since Roughgarden believes that the hetero/homo distinction is a purely cultural creation and not a fact of biology, she thinks it is only a matter of time before we return to the standard primate model. "I'm convinced that in fifty years, the gay-straight dichotomy will dissolve. I think it just takes too much social energy to preserve. All this campy, flamboyant behavior: it's just such hard work."

Despite Roughgarden's long list of peer-reviewed articles in prestigious journals, most evolutionary biologists remain skeptical of her conclusions. For one thing, it's tough to measure the benefits of diversity — or lesbian pair bonding. It's even harder to imagine how traits that are good for the group get passed on by individuals. (As a result, group selection has largely been replaced by kin selection.) In the absence of anything conclusive, most scientists stick with Darwin and Dawkins.

Other biologists think Roughgarden is exaggerating the importance of homosexuality. Invertebrate zoologist Stephen Shuster told *Nature* that Roughgarden "throws out a very healthy baby with some slightly soiled bathwater." And biologist Alison Jolly, in an otherwise positive review of *Evolution's Rainbow* for *Science*, conceded that Roughgarden ultimately fails in her ambition to "revolutionize current biological theories of sexual selection." As far as these mainstream biologists are concerned, Roughgarden's gay primates and transgendered fish are simply interesting sexual deviants, statistical outliers in a world that contains plenty of peacocks. As Paul Z. Myers, a biologist at the University of Minnesota, put it, "I think much of what Roughgarden says is very interesting. But I think she discounts many of the modifications that have been made to sexual selection since Darwin originally proposed it. So in that sense, her Darwin is a straw man. You don't have to dismiss the modern version of sexual selection in order to explain social bonding or homosexuality."

Roughgarden remains defiant. "I think many scientists discount me because of who I am. They assume that I can't be objective, that I've got some bias or hidden LGBT agenda. But I'm just trying to understand the data. At this point, we have thousands of species that deviate from the standard account of Darwinian sexual selection. So we get all these special-case exemptions, and we end up downplaying whatever facts don't fit. The theory is becoming Ptolemaic. It clearly has the trajectory of a hypothesis in trouble."

Roughgarden's cataloging of sexual diversity has challenged a fundamental biological theory. If Darwinian sexual selection — whatever its current variant — is to survive, it must adapt to this new data and come up with convincing explanations for why a host of animals just aren't like peacocks.

MICHAEL D. LEMONICK

Let There Be Light

FROM *Time*

RICHARD ELLIS PACES impatiently back and forth across a small room lined with computer terminals, trying to contain his mounting frustration. The British-born astronomer, now at Caltech, has been granted a single precious night to use one of the twin Keck telescopes, among the most powerful in the world. Last night he and his observing partner, a graduate student named Dan Stark, flew 3,000 miles from Southern California to Hawaii, where the Kecks are located. And during most of the afternoon and early evening today, they've made their final plans for the "run," as astronomers call a night of peering into the heavens.

But things are not going right. It isn't the weather, which is what usually trips up stargazers. Here at Keck headquarters in the sleepy town of Waimea, nestled in the midst of cattle-ranching country on Hawaii's Big Island, thick clouds are scudding past, occasionally dipping low enough to send a driving mist across the grassy hills. But the telescopes are some 25 miles away and more than 2 miles up, in the thin, frigid air at the summit of the extinct volcano Mauna Kea. At an altitude of nearly 14,000 feet, the observatory sits well above the cloud deck. Live video-camera images piped down to the Waimea control room show white domes silhouetted against a fading but crystal-clear sky.

The problem is that Keck 2, the scope Ellis and Stark have been assigned for the night, stubbornly refuses to focus. Time and again, the professional telescope operator who sits in a control room up on the summit and actually runs the mammoth instrument has issued the command that tells it to focus. Time and again, the focusing routine has responded to his commands by crashing. For half an hour engineers have been trying to figure out what is going on — while the first of the precious celestial objects on Ellis and Stark's observing schedule sinks inexorably toward the horizon. "This is pretty profound," says Ellis bitterly. "If you can't focus the telescope, you're stuffed."

No astronomer likes to be cheated out of an observing night, whether the quarry is a mundane moon of Jupiter or an exotic quasar halfway across the cosmos. But Ellis has special cause for frustration: he's looking for something far more elusive than any quasar. Tonight he intended to bag something most astronomers consider next to impossible: the most distant galaxy ever seen — and not the farthest by just a little bit. The current record for distance, held by another giant Mauna Kea observatory, Japan's Subaru telescope, is for a galaxy whose light started its journey to Earth a billion years or so after the Big Bang. But Ellis and Stark suspect they have found not one but six galaxies from an astonishing half a billion years earlier still. Tonight's run could confirm it.

A discovery like that would give Ellis bragging rights at astronomy conferences for years to come, and it would let Stark finish his dissertation with a dramatic flourish. But far more important, it would give astrophysicists their first real glimpse into a crucial and mysterious era in the evolution of the cosmos. Known as the Dark Ages of the universe, it's the 200-million-year period (more or less) after the last flash of light from the Big Bang faded and the first blush of sunlike stars began to appear. What happened during the Dark Ages set the stage for the cosmos we see today, with its billions of magnificent galaxies and everything that they contain — the shimmering gas clouds, the fiery stars, the tiny planets, the mammoth black holes.

When the Dark Ages began, the cosmos was a formless sea of particles; by the time it ended, just a couple hundred million years later, the universe was alight with young stars gathered into nascent galaxies. It was during the Dark Ages that the chemical elements we know so well — carbon, oxygen, nitrogen, and most of the rest — were first forged out of primordial hydrogen and helium. And it was during this time that the great structures of the modern universe — superclusters of thousands of galaxies stretching across millions of light-years — began to assemble.

*

So far, however, even the mightiest telescopes haven't been able to penetrate into that murky era. "We have a photo album of the universe," says Avi Loeb, a theoretical astrophysicist at Harvard University, "but it's missing pages — as though you had pictures of a child as an infant and then as a teenager, with nothing in between."

The full answer may have to wait for a new generation of telescopes expected to come online within the next decade. In astronomy, size matters, especially for faraway objects. The bigger a telescope, the more of a distant galaxy's meager light it can gather — just as a swimming pool catches more rain than a bucket. So astronomers are looking forward to a ground-based monster with nearly ten times the light-gathering area of the Keck, a space telescope more than ten times as big as the Hubble and several radio telescopes with unprecedented sensitivity. Meanwhile, using the basic laws of physics, sophisticated computer simulations, and tantalizing hints from existing telescopes, astronomers have put together a plausible scenario of what must have happened during the Dark Ages.

The first of those hints comes from the universe-wide flash of light that followed nearly half a million years after the Big Bang. Before that flash occurred, according to the widely accepted "standard model" of cosmology, our entire cosmos had swelled from a space smaller than an atom to something 100 billion miles across. It was then a seething maelstrom of matter so hot that subatomic particles trying to form into atoms would have been blasted apart instantly and so dense that light couldn't have traveled more than a short distance before being absorbed. If you could somehow live long enough to look around in such conditions, you would see nothing but brilliant light in all directions.

But as the universe expanded, it finally cooled down enough to allow atoms to form and light to shine out across open space. The accidental discovery of that light back in the 1960s convinced astronomers that the Big Bang was a real event, not just a theoretical construct.

That first detection of the remnants of the Big Bang was crude, but a series of increasingly sophisticated instruments, culminating in the Wilkinson Microwave Anisotropy Probe (WMAP) satellite in 2003, have laid bare the structure of the 400,000-year-old cosmos — only a few hundred thousandths of its present age — in surprising detail. This was the baby picture Loeb referred to. At that point, the universe was still a very simple place. "You can summarize the initial conditions," says Loeb, "on a single sheet of paper." Some regions were a tiny bit denser than average and some a little more sparse. Most of the stuff in it — then and still today — was the mysterious dark matter that nobody has yet identified, largely because it doesn't produce light of any sort. The rest was mostly hydrogen, with a bit of helium mixed in. So far, the universe hadn't done much of anything.

At the start of the Dark Ages, there were no galaxies, no stars, no planets. Even if there had been, we wouldn't be able to spot them. That's because hydrogen-gas clouds are nearly opaque to visible light; no ordinary telescope will ever be able to see what happened afterward. Yet somehow the matter that started as a sea of individual atoms managed to transform itself into something more. So back in the early 1990s, Loeb began lobbying theorists to make a major push to deduce through computer simulations how the first stars formed. The plan was to recreate the young universe digitally, plug in equations for the relevant physics, and see what must have happened.

At first, the simulations agree, gravity was the only force at work. Regions of higher density drew matter to them, becoming denser still — a pattern preserved to this day in the distribution of galaxies, with huge clusters where there were high-density regions back then and great voids in between. Eventually clouds of hydrogen became so dense that their cores ignited with the fires of thermonuclear reactions — the sustained hydrogen-bomb explosions, in essence, that we know as stars. But whereas the familiar stars of the Milky Way are mostly similar in mass to the sun, these first stars were, on average, gigantic — at least twenty-five times as massive as the sun and ranging as much as one hundred times as massive, if not more. A star that big burns very hot, shining perhaps a million times brighter than the sun and generating a wind of particles that pushes the surrounding gases outward, keeping them from collapsing on their own to form new stars. The very first galaxies in the young universe may well have been microgalaxies, as theorist Mike Norman of the University of California at San Diego calls them: each one a single, huge, superhot star, surrounded by a halo of hydrogen.

Because they were so hot, the first stars would have poured out not just visible light but also copious amounts of high-energy ultraviolet radiation. One effect of that radiation would have been to knock apart hydrogen atoms, thus destroying their ability to block light. That process is known as reionization, and those stars, forming perhaps 100 million years into the Dark Ages, or roughly at the era's midpoint, might have rendered the universe transparent on their own if they had lived long enough. But unlike the sun, which has survived 5 billion years so far and should live another 5 billion, those stars lasted only a paltry million years. If the first stars formed 100 million years after the Dark Ages began, they were gone by 101 million years. As they died, the smaller of the stars exploded and spewed their contents back into space, while the bigger ones formed black holes.

As in any other fusion reaction, the fires that powered these short-lived stars worked by forcing simple hydrogen and helium atoms to meld into heavier, more complex elements. The stars that died explosively spiked the surrounding gas clouds with elements like oxygen and carbon, which had never existed before. Billions of years later, the elements forged in stars like these would be assembled into planets, organic molecules, and, ultimately, human beings. At the time, though, they served simply to change the chemistry of the clouds, allowing them to collapse into far smaller objects than they could before. The second generation of stars, incorporating the ashes of the first, arose almost immediately. They were much more like the sun, in both composition and size. And like the sun, they would have started out generating lots of ultraviolet light before settling down to a more sedate existence.

It's this radiation — the ultraviolet light from hot, newly formed stars — that many theorists suspect finally reionized the remaining hydrogen, making it transparent again and bringing the Dark Ages to a close. Others suggest that the process may have been powered instead by black holes spewing out X-rays and ultraviolet light. Or it may have been a combination of hot stars and black holes that cleared the hydrogen and put an end to the Dark Ages.

In any event, the accepted scenario is that this new generation of small galaxies, containing no more than a million second-generation stars, gradually collided, merging to form ever bigger objects that eventually reached the size of the Milky Way. One piece of evidence: the faintest and oldest galaxies found in any great number by the Hubble telescope tend to be small and irregular in shape, not the majestic spirals and huge elliptical galaxies that formed later. Another hint that the merger theory is correct is that the collisions are still going on today. Astronomers can see hundreds of colliding galaxies in their telescopes, and our own Milky Way is still slowly gobbling up the half-dozen or so dwarf galaxies that surround it.

Guessing how the very first stars formed is relatively easy, since the universe was so simple at the start of the Dark Ages. By the end, however, things were starting to get complicated. The stars had begun to affect their environment, and the environment in turn affected the stars in feedback loops that nobody has completely figured out. That's why astronomers want to test their theories with observation, and they will need a new generation of telescopes to do so.

To spot the earliest objects, however, astronomers will have to stop looking for ordinary light. The universe has expanded vastly since its earliest days — but it isn't that galaxies and other objects are flying apart. Rather, it's that space itself has been stretching — a difficult concept even for a physicist to grasp, but which must be true according to the equations of relativity. Cosmologists say you should imagine the universe as a balloon with dots painted on its surface. As the balloon inflates, the dots will get farther apart — not because they're sliding around but because the balloon is stretching.

That being the case, a light beam traveling through expanding space is stretched as well, its wavelength getting longer as it goes. Long-wavelength light is red; stretch it out longer and it becomes infrared light and then microwaves and, finally, long-wavelength radio waves. The flash that came from the Big Bang started out as visible light; by now, 13.7 billion years later, it's still streaming through space, but it has been stretched so much that astronomers have to use microwave antennas to detect it. The earliest galaxies came after the Big Bang, so their light isn't quite as old, hasn't been traveling as long, and thus isn't stretched as much. That light should be detectable not as microwaves but as infrared — which is why the new telescopes will be fitted with infrared sensors. It's also why the James Webb telescope, NASA's planned successor to the Hubble, will be optimized to see infrared, not visible light.

Still other telescopes will be trying to take pictures not of the first stars and galaxies but of the clouds of hydrogen atoms from which they formed and that they eventually destroyed. The hydrogen atoms emitted radiation too, in the form of radio waves, and several competing projects in various stages of completion in India, China, the Netherlands, and Australia are being designed to see them. The last, known as the Mileura Widefield Array, is considered the most promising because its five hundred separate antennas will be located on a remote cattle station in Western Australia, far from any interference from earthly radio broadcasts. "The South Pole would be good too," says Jacqueline Hewitt, director of the Institute for Astrophysics and Space Research at MIT, which is a partner in the project, "but this is a bit more accessible. We'll need to cut some roads, though."

What makes Mileura and the other projects so powerful is that by tuning the receivers to different radio frequencies, they will be able to pick up signals broadcast by hydrogen atoms at different periods in the Dark Ages. When you map cosmic hydrogen at, say, 50 million years after the Big Bang — before the first stars had a chance to form — then at 100 million, 200 million, or half a billion years later, you get a series of snapshots. Combine them, says Loeb, and "you'll be able to make a 3-D picture of hydrogen gas as the universe evolves. At some point, you'll start to see holes, like Swiss cheese," as the gas clouds become ionized and transparent. Precisely how the holes grow and merge over time will help determine whether the clearing out is being done by small galaxies, big black holes, or something entirely different — and depending on which it is, some theorists could be vindicated and others refuted. But astronomers will at last have an answer to the mystery they have puzzled over for a decade and a half.

In the meantime, observers have been chipping away at this mystery as best they can with the tools they have. One important clue comes from the observation of the most distant quasars — objects believed to be giant black holes swallowing huge volumes of gas at the cores of young galaxies. The Sloan Digital Sky Survey, a comprehensive scan of the heavens, has turned up several of these from about a billion years after the Big Bang. By watching how the light of quasars is altered by surrounding gas, astronomers have concluded that there was still some atomic hydrogen around then, although not much.

But the WMAP satellite, launched to look for light left from the Big Bang using a broadly analogous technique, determined that the clearing out of hydrogen between the stars was well under way much earlier, just half a billion years post–Big Bang. "Theorists have been telling us that it should have happened fairly quickly once it began," says Michael Strauss, a Princeton University astronomer and deputy project scientist for the Sloan survey. "But the observations may be telling us otherwise."

Indeed, observations often take theorists by surprise. Last fall a focus on one tiny region of the universe by the Hubble, the Spitzer space telescope (which operates in infrared wavelengths), and the European Space Organization's ground-based Very Large Telescope in Chile revealed the existence of a galaxy dating to about 1 billion years after the Big Bang that was far larger and more mature-looking than the primordial dwarf galaxies everyone assumed they would see. "It was unexpected," admits Mark Dickinson of the National Optical Astronomy Observatory, in Tucson, Arizona, who worked on the project. "But maybe it shouldn't have been." The theorists might have things all wrong. But it could also simply be that any population will have a few individuals that are way outside the average — humans who stand over 7 feet tall, for example. They're very noticeable but not at all typical.

Until someone finds better evidence to the contrary, it's safe to assume that the very tiny galaxies filled with second-generation stars were by far the dominant type in the early cosmos. It would also have been safe to assume that nobody could spot them in their earliest incarnation without giant new telescopes — if not for Ellis. "He does like to push the frontiers," says theorist Norman with mixed amusement and respect. "It's always great fun to go to a meeting and see the latest Ellis most-distant-object sweepstakes entry."

Ordinarily, Ellis explains, you could never see small galaxies a mere 500 million years after the Big Bang; they're just too faint for any telescope now in existence. But the universe itself has supplied a way of boosting a telescope's magnifying power. The theory of relativity says massive objects warp the space around them, diverting light rays from their original path. In the 1930s Albert Einstein realized this meant a star, say, could act as a lens, distorting and amplifying the light from something behind it. In practice, he said, it probably happens so rarely that we will never see it.

Einstein was wrong. So-called gravitational lenses have become a major factor in modern astronomy. They have revealed, among other things, the existence of tiny planets around stars thousands of light-years away and have created weird optical effects, including multiple images of faraway quasars. If you look at a massive cluster of galaxies, Ellis figured, you might see amplified images of more distant galaxies, too faint to be seen otherwise. So a year or two ago he started aiming the Keck at galactic clusters, and along with Stark, he identified six candidate objects. To make certain that these were truly far away, the pair has come back to the Keck for a second, more intensive look. "We want to be absolutely sure we aren't fooling ourselves," says Ellis. "Before we claim we've really found them."

For an hour or so, it looked as though they wouldn't get the chance. They had just this one night at the Keck; the telescope is so overbooked that even an eminent astronomer like Ellis has to wait his turn, and his next observing run isn't until January 2007. But the engineers this night have figured out the problem. When Stark entered his user name in the online telescope log, he made a typo. Every time the focusing routine came upon it, the program froze. The typo has now been corrected. The Keck can focus again, and to their delight, Stark and Ellis are able to confirm that at least three of their faint galaxies do seem to lie hundreds of millions of light-years farther away — and hundreds of millions of years closer to the Big Bang — than anything ever seen before.

"It's regrettable that we couldn't check the other three," Ellis says a few days later, "but we're now very confident and very excited. If we've found this many in such a tiny area of the sky, there could be enough of these small galaxies to supply a substantial fraction of the energy that reionized the universe. I'm very confident that we have an important result."

If anything, that's an understatement. The first galaxies to emerge from the blackness of the early universe can't be studied in detail until telescope technology makes another great leap. But Ellis and Stark may have got a glimpse — and given theorists the first hard evidence — of that unimaginably distant time when the cosmos left infancy behind and entered the formative childhood that led, eventually, to our sun and the tiny blue planet that circles it.

JEFFREY A. LOCKWOOD

The Nature of Violence

FROM *Orion*

I HAVE BEEN ATTACKED by animals for thirty years. Working at a veterinary clinic in high school, I learned the skill of keeping snarling dogs at bay with a squeegee and the art of restraining injured cats. Later, in college research laboratories, I encountered the occasional frightened rat willing to use its yellowed incisors in self-defense. The mice seemed less threatening, but were quicker to draw blood if carelessly handled.

But I was unprepared for the unmitigated ferocity of the Gryllacrididae — insects that look like a cross between a cricket and a grasshopper. My research as a professor at the University of Wyoming had focused on grasshoppers, and I had the opportunity to spend a sabbatical leave in Australia working with their relatives. While grasshoppers can be feisty, kicking and struggling valiantly when held, gryllacridids are fierce. Lifting their wings to appear larger and thumping their abdomen against the ground like a war drum, they launch themselves at any intruder. Their gaping, sickle-shaped mandibles leave no doubt that they intend to inflict as much damage as possible, and the largest species can leave a deep gash in a person's hand.

Gryllacridids attack in order to live. The smallest of these insects are no larger than a pencil stub, while the largest species are meatier than a man's thumb. These creatures lack standard insect defenses — they have no stench, toxin, or sting. Given their bite-sized morphology and their lack of chemical protection, gryllacridids make tempting hors d'oeuvres for various predators. The insects' primary defense is a set of characteristics that evolved to help them avoid such a fate by escaping detection. In coloring they tend toward drab, earthy shades. A row of pinned specimens in a museum drawer might well match a standard selection of men's shoes. Their nests, whether subterranean burrows or mats of dried leaves, are invisible during the light of day. And finally, gryllacridids are strictly nocturnal, avoiding the visually oriented birds that constitute the primary predators of many insects. There are, however, plenty of nocturnal hunters seeking a midnight snack. So, when stealth fails and a gryllacridid perceives that it is faced with imminent death, it unleashes a maniacal display of mandible-gnashing, abdomen-thumping, wing-flapping ferocity. These savage creatures gave me ample opportunity to ponder the nature of violence.

Starting a new line of study in a novel place reminded me of my first days of graduate school a dozen years earlier. Dr. Jeff LaFage taught my graduate course in insect behavior at Louisiana State University with passion and intensity. He was a demanding and kind teacher, a true scholar and Renaissance man — studying the evolution of termite sociality, collecting Tiffany glass, and hosting a Baroque music program on public radio. He was unable to harm a fly without reason, but when he peered over the wire rims of glasses perched menacingly on his nose, a student who had failed to read the assigned paper or otherwise prepare for the semiweekly grilling cringed. Loathing pretense, he insisted on being called Jeff rather than Dr. LaFage, but he was one of the few academics who evinced the original meaning of the honorific title, meaning "to teach." And so in my mind he remains Dr. LaFage.

Through the fall of 1983, Dr. LaFage revealed to us the ways of insects: the wonders of territoriality, the characteristics of communication, the qualities of sociality, and the origins of aggression. He would often use imaginative ecological and evolutionary scenarios to set up a series of Socratic questions that would elicit from us the answers about the complex origins of behavior.

Behavior, Dr. LaFage explained, is a function of how an animal perceives itself and the environment, as constrained by evolution and experience. Insects rely heavily on instinct, which serves them well. For example, many insects respond to a sudden surge of carbon dioxide, a sure indication that a large mammal is nearby. For mosquitoes this gas is a chemical dinner bell. But a huff of breath on a flower head induces thrips to hastily abandon their refuge.

Like a swarm of living commas, they evacuate rather than risk being consumed by a lumbering grazer oblivious to their presence. Similarly, many social insects such as bees and ants rush from their nests when carbon dioxide pours in, but theirs is a charge rather than a retreat: the soldiers deploy to attack whatever creature has its face close enough to generate the chemical alarm.

For a man of such a gentle, albeit intense, demeanor, Dr. LaFage seemed to relish his review of the insectan arsenal. These creatures have mouthparts variously adapted for crushing, dismembering, slicing, and piercing. The most remarkable structures are those whose original function was co-opted for assault. In colonies of ants, bees, and wasps, only the queen reproduces; the other females have no use of their reproductive tracts. In this celibate caste, structures for depositing eggs gave way to stinging barbs, and cells that formerly produced lubricating secretions during egg-laying became poison glands. Organs once used for perpetuating life evolved into structures for killing.

But social species with the capacity to kill one another have also evolved sophisticated means of communicating relatedness and submission. Ants, bees, and termites identify nest mates by chemical cues. Only odd-smelling intruders are quickly set upon and stung mercilessly or torn to shreds. Among nonsocial insects, there are no constraints on killing. If a hungry grasshopper stumbles upon a recently molted comrade, it will devour the helpless creature.

Dr. LaFage made clear that the balance sheet of evolution guaranteed that the genetic or energetic cost of violent behavior was inevitably offset by the potential benefits that it might yield. The ferocity with which the bees defend their hive arises from a genetic quirk ensuring that the workers are all intimately related sisters, sharing more genetic similarity with one another than with their mother. As a consequence, their willingness to die for the collective ensures a continued production of sisters and thereby genetic facsimiles of themselves.

Dr. LaFage was a bioeconomic hardliner. The currency of evolution is the gene: the more copies of you, the richer you are. When he spoke of his daughters, the tender tone belied his evolutionary dogmatism, but when it came to insects, he maintained strict scientific objectivity. He brooked no sympathy for the worker bee eviscerating herself as the inevitable result of stinging to defend her colony. This suicidal creature was simply placing an unconsciously calculated bet with a well-established genetic payoff — if the queen is killed, then the biological factory producing more sisters is destroyed. For most creatures, only one's own life takes on this value, which explains — at least in part — why predators have only a 10 percent success in securing a meal following an attack. If the predator fails, it goes hungry. If the prey fails, it dies. Dr. LaFage rattled off a litany of extreme, last-ditch efforts used by insects to escape the grip of their predators: beetles that squeak or bleed spontaneously, moths that flash hind wings with owl-like eyespots, and crane flies that sacrifice still-twitching legs. When the cost is your life, whether corporeal or genetic, the nothing-to-lose approach becomes viable.

Gryllacridids are extremely difficult to observe in their natural habitats. I spent hours in the bush searching for these creatures with a headlamp, under the tutelage of one of Australia's finest entomologists. Even with his instructions, I failed to see a single gryllacridid in the wild. More recently, a graduate student in Western Australia managed to track individuals of two species over several nights. His dissertation documented the complex homing behaviors of these insects, including their capacity to use dimly illuminated features of the landscape as navigational beacons. But he did not observe any encounters between individuals. It seems that gryllacridids are fiercely territorial, having evolved the rudiments of a primitive social system that minimizes the risk of conflicts through the avoidance of their own kind.

While the young may aggregate in tight groups, presumably to pool their defenses against larger predators, adult gryllacridids lead isolated lives, seeking one another's company only for mating. The solitary adults forage in territories carved from the harsh Australian outback. They exhibit a range of feeding habits but may be typified as nocturnal scavengers and opportunistic predators. The formidable mandibles of these insects are well adapted for crushing seeds and dismantling exoskeletons — including those of their own relatives. The violence of these reclusive insects is bounded only by their physical capacity to inflict harm. In the laboratory, gryllacridids must be housed separately to prevent them from killing each other.

Part of my laboratory research in Australia involved collecting the silk that the gryllacridids produce to line their burrows or bind their leafy nests. In the wild, the smell of the silk enables an individual to relocate its nest at the end of a night's foraging. In the lab, each insect was provided with an index card "tent" that served as a nest-building site. Within twenty-four hours, it had constructed a rudimentary matrix of snow-white silk extruded from its salivary glands, and with a second night's work, it had woven a dense tangle of threads to impede the intrusion of predators. Before collecting this silk for chemical analysis, I used a thin glass probe to harass the tent-maker into abandoning its nest, sometimes resorting to gently shoving recalcitrant insects out of their homes. In response to disturbance, one individual drummed its abdomen against the floor of the tent in a threatening display of impending violence. When I tried pushing this creature from its nest, the insect crushed the probe between its mandibles — the equivalent of snapping a matchstick.

My laboratory confrontations mimicked predator-prey encounters. The insect's initial strategy upon being discovered was to retreat into its burrow or nest. Only with continued provocation or if flushed from its refuge did it switch from retreat to attack. Some species of gryllacridids had a very high threshold for aggression, resorting to attack only when driven into the open. Others rapidly switched from defensive withdrawal to unabashed assault.

Once engaged, the gryllacridids pressed the attack with reckless abandon. For these creatures, there was no halfhearted display of aggression, no gradual escalation of hostility. Moreover, they appeared utterly incapable of tempering their rage in proportion to the size of the intruder and the corresponding likelihood that such ferocity would be effective. While the largest species could readily draw blood, the smallest could not manage to pinch a bit of skin between their mandibles. Nonetheless, a cricket-sized individual would posture menacingly and snap at any object within range, even if it was a thousand times bigger than the insect itself. When a large individual became aggressive, the only way to handle it was to place the entire cage in the refrigerator for fifteen minutes. When the insect was immobilized in cold-stupor, I could move it into a fresh cage.

After months of working with gryllacridids, I came to understand some of what it means to be such a creature. I valued these insects as marvels of evolution — exemplars of wildness and ferocity — but also as teachers of life's harsh lessons. How the will to live becomes a willingness to kill was a phenomenon I encountered daily. I couldn't say that if I were a gryllacridid I'd have acted any differently. I could say, however, that I regretted that they were unable to habituate to me. Their fear made both our lives more difficult.

One morning not long before I was to leave Australia, I injured one of the larger gryllacridids. The creature had been engaged in a wildly aggressive display, simultaneously flaring its wings and gnashing its mandibles. In the course of this frenzy, it had managed to climb to the edge of the cage to press its attack. Setting the lid down too fast, I accidentally pinned the insect against the top edge of the cage, where it thrashed and twisted. The force of the lid ruptured the tender abdominal membrane. The insect fell back into its container, and when the cage lid effected my disappearance from the scene, it calmly assessed the situation. A globule of yellow-streaked fat oozed through the gaping wound. It then curled its head down toward the leaking viscera and proceeded to consume its own entrails.

My heart sank — and with this I knew I'd violated the scientist's dictum of objectivity. I loved these wild animals, not with the sort of conditional affection that we have for organisms that can return our warmth but with a sort of deep empathy without pity. For we ultimately shared a defining reality: the capacity for animate relationality — for striking out in fear, attacking in anger, and writhing in pain. I believed that I finally understood what Walt Whitman meant in describing animals in *Leaves of Grass:* "So they show their relations to me and I accept them / They bring me tokens of myself, they evince them plainly in their possession."

As the gryllacridid cannibalized itself, I tried to rationalize away my feeling of empathetic horror. Physiologists maintain that while insects may be pained by acute pressure, once the body wall is ruptured there appears to be no persistent tenderness. So it was that Dr. LaFage admonished us to avoid subjectivity — not all organisms process sensations in the same way. Their world is not ours. Pain, he said, is cognitive experience, not merely sensation. We cannot know what another animal experiences. Dr. LaFage would not countenance any latent anthropomorphic tendencies in his students. My feeling about what I'd done to the mangled gryllacridid did not reflect my training as a researcher. Nor did it mirror the insect's likely reaction were I the one who had been crushed. Rather, my response was essentially that of a poetic scientist, a compassionate human. As empathetic creatures, we cannot stop ourselves from imagining the pain of other sentient beings.

The year after I left Louisiana and came to Wyoming as a freshly minted Ph.D., Dr. LaFage entertained a visiting scientist in the French Quarter of New Orleans. He was escorting this female colleague when a mugger grabbed her purse. The woman became tangled in the strap, and Dr. LaFage stepped between them and said, "Don't hurt her, you can have the purse." I can picture him doing this, with the soothing confidence of a man who knows the nature of violence. I have even imagined the setting: the grimy street littered with the refuse of bars and clubs contrasting with elegant ironwork on the balconies overhead; the sultry heat hanging limp like Spanish moss; the blended odors of wetness — sweat trickling down the small of a back, fetid water pooled in gutters, and a whiff of urine from an alley. What I can't imagine is the next moment: the young man drew a gun and fired point-blank.

Dr. LaFage's family and friends struggled to understand this seemingly senseless act. I had no special access into the mind of the killer, who was, I suspect, much like many of the hopeless, poor, angry inner-city youths of our country. But my teacher had given his students a means of making sense of behavior. He had offered us ways of seeing into the lives and conditions of other beings, so that we might begin to understand their aggression. We had learned that violence is the baseline strategy for most encounters between, and indeed within, species. This tendency is reduced only when there is a more successful adaptation to defending oneself or acquiring vital resources.

For most humans, the essentials of life — food, shelter, clothing, self-worth — can be had through nonviolent means. I have no doubt that this gentle and generous man would have given whatever material resource his assailant demanded. But what that angry, scared youth needed could not be given at the point of a gun, in a fleeting moment, on a New Orleans sidewalk. For the essentials of human life are not limited to bodily needs. And robbery is not always about material gain. Human cultural evolution has produced complex rules to constrain instinctual violence, but these norms must be learned within a viable social network. Even so, the development of human technology has outpaced behavioral adaptations. Our weapons enable us to draw and fire a gun much faster than another human can submit to our rage or plead for mercy.

So it was that Dr. LaFage's teachings proved both vitally important and ultimately inadequate to understanding the genesis of human violence. I cannot weigh the despair of an inner-city teen; I have no formula that predicts when a fellow creature's fear turns to rage. Nor can I graph a student's affection for a gentle, demanding mentor. In the end, of course, Dr. LaFage was right: we cannot truly know the pain of others. Perhaps it is good that we cannot fully share in the anguish of terrified animals, dying men, grieving widows, fatherless children, or soulless youths. Sometimes our own sadness is as much as we can bear.

LYNN MARGULIS AND EMILY CASE

The Germs of Life

FROM *Orion*

WATCH TV for an hour. Flip through a mainstream magazine. Peruse personal hygiene or cleaning products in a store. You'll feel the need to defend yourself with antibacterial soaps and cleaning agents, even antimicrobial pillows and socks. Fear of bacteria has reached a feverish pitch recently, thanks in large part to the work of ever-industrious advertisers.

In our efforts to eliminate these "germs" we have had devastating effects — not on the bacteria, but on ourselves. The bacteria that now pose the greatest threats to humans are products of our own making. The evolution of pests and pathogens resistant to human poisons has a long, well-documented history. Hospitals, where antibacterial drugs, soaps, and cleaners are used in volume, are hotbeds of antibiotic-resistant strains of bacteria. Farmers feed livestock excessive amounts of antibiotics, thereby selecting for bacteria that are resistant to those medicines — versions of which are also used for humans.

But our xenophobia also blinds us to a more fundamental insight: the health of our environment, and our bodies, depends on bacterial communities. Indeed, they are responsible, as ancestors, for our very existence.

If Life had a yearbook, bacteria would win all of the awards, especially "most likely to succeed." A bacterium is an organism made up of one or more small prokaryotic cells, those that have DNA genes but lack nuclei and chromosomes. Bacteria inhabit the farthest reaches of the biosphere. They live in the hottest, coldest, deepest, saltiest, and most acidic environments. They are the most ancient life form, having lived on Earth for at least 3.8 billion years, over 80 percent of its history. By contrast, humans have occupied a narrow range of environmental conditions — and for only about 0.003 percent of the Earth's existence. If we even made it into the yearbook, the caption would have read "photo not available."

Earth's environment is in large part the product of bacterial metabolism. Bacterial nitrogen fixation enriches the soil at no cost to us. And the photosynthesis that excretes oxygen and makes food for all life is carried out by the blue-green bacteria called cyanobacteria — both the free-living kind and those that became chloroplasts in the cells of algae and plants. These are just two of bacteria's life-sustaining processes, invented at least 2 billion years ago. We should view them as the wisdom of the ancients.

Even disease-causing bacteria — exceedingly rare despite the fear-mongering of marketers — play a part in ecological health. Anthrax spores, for example, float in the dust of overeaten and sun-exposed fields, enter the lungs and blood of vulnerable or weak grazers, and kill them. Fields recover their vegetation. The grazers' food supply is spared, the stability of the ecosystem restored.

Bacteria also sustain us on a very local, intimate scale. They produce necessary vitamins inside our guts. Babies rely on milk, food, and finger-sucking to populate their intestines with bacteria essential for healthy digestion. And microbial communities thrive in the external orifices (mouth, ears, anus, vagina) of mammals, in ways that enhance metabolism, block opportunistic infection, ensure stable digestive patterns, maintain healthy immune systems, and accelerate healing after injury. When these communities are depleted, as might occur from the use of antibacterial soap, mouthwash, or douching, certain potentially pathogenic fungi — like candida or vaginal yeast disorders — can begin to grow profusely on our dead and dying cells. Self-centered antiseptic paranoia, not the bacteria, is our enemy here.

But in our ignorance, we also miss a larger lesson. Bacteria offer us evidence that health depends on community, and independence is an ecological impossibility. Whenever we treat isolated medical symptoms or live socially or physically isolated lives, we ignore warnings from our more successful planetmates.

Bacteria in their natural environments live in well-structured communities based on reciprocity. As one type excretes acid, sugar, or oxygen, its wastes become food or gas for others. And these communities are ecologically sensitive. Bacteria change form and metabolism in response to environmental cues like dryness or heat. Many multicellular bacteria (such as those made of long filaments of cells) revert to single cells in the laboratory. But in the richness of their normal habitat, from pond water to tongues, they transform back into their long chains.

The bacterial propensity to live in ecological communities has also left its mark in the cells of all larger life. Protoctists (like algae and ciliates) and fungi (like yeasts and molds) — not to mention plants and animals — are all nucleated-cell organisms; their cells contain nuclei that divide by mitosis, a complex dance of chromosomes. As research from our lab and others has proved, nucleated-cell organisms could not have evolved without the multimillion-year-old permanent mergers of specific bacteria. Cellular respiration, for example, the process that releases energy from food, occurs in the cell's mitochondria. Mitochondria were once independent bacteria that attacked, or were engulfed by, an early protist.

More recently, some of us have studied what we think is another historic incorporation of bacteria. This one involves the wily bacteria known as spirochetes, including one that we suspect is an ancestor of all of us nucleated-cell organisms. By new molecular biology techniques we expect to prove that an ancient spirochete fused with another very different bacterium, and that the result was that certain free-swimming spirochetes contributed remnants of their lithe, snaky bodies to become moving components of cells. These parts include the familiar waving hairs called cilia and the tubules of the mitotic spindle, which moves chromosomes so that cells divide equally.

But an even later consequence of the hypothetical merger evidently extends to sensory tissues. In mammals, the cells of the tongue's taste buds, the inner-ear cells required for hearing, and light-sensitive cells in the retina of the eye all have traceable, peculiar features in common. Even cells of the semicircular "canal-balance organ," the stimulus-receiver that tells us whether we are on our feet or upside down, share the detailed features we interpret as clues to their origin.

The salient feature is that these cells have the hairlike cilia, which sense stimuli like light, touch, and sound. It is widely accepted that these cilia, all composed of skinny tubules arranged in a distinct pattern, evolved from a common ancestor, whose identity remains unknown. Our evidence indicates that it was the ancient spirochete: that in the complex ecology of bacterial communities, the merger happened; and that ultimately out of that merger our sensory apparatus evolved, giving us the basis of our awareness — and by extension our consciousness.

In the symbiotic associations that have persisted, cohabitation ultimately succeeded. Our nucleated-cell ancestors evolved because they could swim, breathe oxygen, eat whole bacteria, and merge. Their success was predicated on an attraction to sugars and each other, struggle, fusion, eventual incorporation, and integration by compromise. Our sensibilities come directly from the world of bacteria. Like all life, we thrive in communities. It's natural that people who have strong social relations prove healthier and longer-lived.

Humans have nonetheless found no shortage of ways to foul communities, cause extinctions, and threaten our own existence in the process. But bacteria wouldn't miss us. They have run the planet for most of its history, and our rush to kill them indiscriminately reveals only our own naiveté. The bacteria, with their complex history and virtuoso performances in energy and food recycling, will easily endure our assault. But our own survival depends on a revolution in human attitudes toward — and ability to learn from — our microbial ancestors.

STEVE OLSON

Neanderthal Man

FROM *Smithsonian*

As a boy in Sweden, Svante Paabo read everything he could about ancient civilizations. After powerful North Sea storms uprooted trees, he begged his parents to take him to archaeological sites to look for potsherds and other artifacts. When he was thirteen, his mother, a food chemist in Stockholm, yielded to her son's most frequent request: to visit Egypt. "It was absolutely fascinating," he recalls. "We went to the pyramids, to Karnak and the Valley of the Kings. The soil was full of artifacts."

Paabo, fifty-one, is still looking for artifacts, but in a very different place. He's a leader of the worldwide quest to explore the past by analyzing human DNA. He has helped show that human groups — southern Africans, Western Europeans, Native Americans — are closely related, despite superficial distinctions. He has been uncovering key genetic changes that helped transform our shambling, hirsute ancestors into the brainy bipeds we are today. This past summer, Paabo announced that he and his coworkers were going to take the next — and biggest — step, in their effort to resurrect the genome of the Neanderthal, our distant evolutionary cousin, who went extinct 30,000 years ago. The first scientist to analyze segments of DNA from Neanderthal bones, Paabo now wants to recreate the entire DNA sequence of a Neanderthal and compare it with our own, looking for the reasons that one evolutionary experiment failed and the other succeeded. "He really is a visionary," says Mary-Claire King, a geneticist at the University of Washington.

Paabo is director of the genetics department at the gleaming new Max Planck Institute for Evolutionary Anthropology in Leipzig, Germany. But you'd never guess his heady position from his taste in clothes, which leans toward shorts and Hawaiian shirts. In his simply decorated office, he kicks off his clogs, folds his long legs under his angular body to perch on a sofa, and grins. "It is a wonderful time to be working in this field," he says.

Ever since the 1940s, when DNA was identified as the molecule that carries genetic information between generations, scientists have predicted that the study of genetics would yield great things, from drought-resistant crops to cures for genetic diseases. But more recently geneticists have realized that there is another way of looking at DNA — as a link to history. All of us inherited our DNA from our biological parents, who inherited it from their biological parents, and so on. Like an ancient manuscript that is copied and recopied with each generation, DNA bears tales from beyond memory. Furthermore, scientists can date genetic changes that have occurred by comparing DNA among humans or between humans and other species. In this way, DNA connects us not only with our ancestors but also with the animals from which we evolved.

Paabo enrolled at the University of Uppsala, in 1975, to study Egyptology. But rather than excavate exotic archaeological sites, as he expected, he spent most of his time conjugating ancient Egyptian verbs. "It was not at all what I wanted to do." Soon he found himself in medical school, a route his biochemist father had also taken. Then he entered a Ph.D. program in molecular immunology.

Still, he couldn't shake his fascination with Egypt. "I knew about these thousands of mummies that were around in museums," he recalls, "so I started to experiment with extracting DNA." With the help of his old Egyptology professors, Paabo obtained skin and bone samples from twenty-three mummies. Working nights and weekends (Paabo was worried that his immunology professor would not approve of the project), he succeeded in extracting and analyzing a short segment of DNA from the 2,400-year-old mummy of an infant boy. In early 1985, he sent his results to *Nature,* one of the world's leading scientific journals, which made the paper its cover story — the equivalent in science of hitting a grand slam in your first professional at-bat.

Paabo also sent a copy of the manuscript to Allan Wilson, a molecular biologist at the University of California at Berkeley. Wilson had made headlines when he and his colleagues extracted a fragment of DNA from the remains of a quagga, a zebralike creature that went extinct in 1883. After Wilson read Paabo's paper, he asked if he could go to Paabo's lab for a sabbatical. "I hadn't even finished my Ph.D.!" Paabo says. Paabo wrote back with a counteroffer: Could he work in Wilson's lab?

Wilson, who died of leukemia in 1991 at the age of fifty-six, "was one of the best people I've ever seen at generating ideas," says Mark Stoneking, who worked with Wilson in the 1980s and is now one of Paabo's colleagues at the institute. Stoneking helped Wilson establish the existence of "mitochondrial Eve" — a woman who lived in Africa about 200,000 years ago. The Berkeley scientists traced our ancestry to her by analyzing the DNA in mitochondria, parts of a cell that produce energy and operate independently of the rest of the cell. We inherit the DNA in our mitochondria through our mothers, grandmothers, great-grandmothers, and so on. By analyzing the mitochondrial DNA of people throughout the world, Wilson and his colleagues determined that the maternal lineages of everyone alive today converge on a single ancient woman.

Paabo, meanwhile, was developing new ways of extracting DNA from preserved specimens of extinct organisms, including moas (a giant flightless bird) and marsupial wolves. Others in Wilson's lab were trying to find DNA in fossilized plants and animals. In the acknowledgments of his 1990 novel *Jurassic Park,* author Michael Crichton gives part of the credit for his inspiration to Berkeley's Extinct DNA Study Group.

Paabo landed his first academic post at the University of Munich in 1990. There he expanded his work on the DNA of ancient animals and plants — mammoths, maize, European cave bears. He also resumed his work on ancient human DNA; for example, he was part of the team that managed to sequence some mitochondrial DNA from the "Ice Man," who was frozen into a glacier in the Tyrolean Alps more than five millennia ago and discovered in 1991. That success fired Paabo's ambition to take on one of the toughest questions in paleoanthropology: What is the nature of our kinship with extinct hominids?

In 1856 two quarrymen dug up a set of odd-looking human bones in the Neander Valley, near Düsseldorf, Germany. The remains were the first recognized traces of a group that came to be known as the Neanderthals (*thal* means "valley" in German). For the past 150 years, scientists have argued about the relationship between today's humans and these vanished people. When anatomically modern humans — the ancestors of today's Europeans — began migrating into Europe about 40,000 years ago, did the Neanderthals simply die out? Or did they interbreed with the newcomers, contributing some DNA to the gene pool of today's humans?

Paabo decided to look for DNA in the original Neanderthal bones. Needless to say, the curators at the Rhineland Museum in Bonn, who are responsible for the fossilized bones, were not eager to let him take samples, since analyzing the bones would mean grinding up irreplaceable fossil material. But Paabo persisted, and the curators finally agreed.

A bone specialist sawed a half-inch chunk from the upper right arm bone of a 42,000-year-old Neanderthal fossil. Paabo handed over the sample to graduate student Matthias Krings, who wasn't optimistic — extracting DNA from 3,000-year-old mummies had been hard enough. But soon Krings began to find mitochondrial DNA sequences that were clearly different from those of any human beings living today.

The results, along with those of subsequent studies, indicated that Neanderthals contributed little, if any, DNA to modern humans. Instead, they appear to have been displaced by modern humans — the taller, more graceful creatures with round skulls and prominent chins who first appear in the fossil record in eastern Africa about 200,000 years ago. Paabo's work means that during the thousands of years that Neanderthals shared the continent with modern humans, there was probably little interbreeding between the two groups. The same thing happened in other parts of the world: archaic populations of humans in Africa and Asia gradually went extinct without leaving an obvious genetic trace.

The apparent lack of interbreeding between archaic and modern humans means that we are a very young species — brash upstarts that overran the older and more established species of humans. "In a sense, we are all Africans, though some of us have gone to live in exile," Paabo says. To be sure, physical appearances changed as groups of modern humans moved into different environments. For example, as they moved into northern climates, natural selection appears to have favored lighter skin colors —

probably because lighter skin admits more sunlight and thereby allows the body to synthesize sufficient vitamin D to endure long, dark winters. As a result, over many generations, the occupants of northern Europe and Asia gradually developed lighter skin than their ancestors. But these superficial differences disguise a remarkable genetic similarity. "Different subgroups of chimpanzees, such as those in eastern or western Africa," says Paabo, "have a much longer history of genetic separation than do, say, Chinese and Africans."

The German government provided very little support for anthropological research after World War II, a response to abhorrent wartime activities of the Kaiser Wilhelm Institute for Anthropology, Human Genetics, and Eugenics in Berlin. (The head of the institute supported Nazi racial policies, and his assistant, Josef Mengele, sent body parts from Auschwitz to be studied at the institute.) But following the 1990 reunification of Germany, officials began looking for neglected areas of science to support in the effort to build new ties between East and West. In 1997 the government invited Paabo to move to Leipzig, a university town in the former East Germany, to start a new institute on human evolution with three other prominent scientists: Christophe Boesch of the University of Basel in Switzerland, who studies wild chimpanzees; Bernard Comrie, a linguist from the University of Southern California in Los Angeles; and psychologist Michael Tomasello from the Yerkes Primate Center in Atlanta. In the summer of 1997, the four scientists set off for a hike in the Alps south of Munich to mull over the invitation. By the time they returned from the mountains, they had decided to accept it. "There's no reason to let Hitler keep us from working on human origins anymore," says Paabo.

Originally housed in an old Leipzig publishing house, the Institute for Evolutionary Anthropology moved in 2002 into a new $30 million building south of downtown. The four directors collaborated on the design, with Paabo insisting that a four-story rock-climbing wall be installed in the lobby. The directors agreed to focus their efforts on one particular question: What makes human beings unique? And to avoid empty speculation, they decided to work only on questions for which data are available. "The kinds of questions we ask are ones where we can see how to go about finding answers," says Comrie.

One day Tomasello and Paabo were talking in the institute's cafeteria about a family in England with a remarkable genetic defect. Some members of the family have a mutation in a gene known as FOXP2, which helps direct the development of the brain during infancy and childhood, and every family member with the mutation had great difficulty speaking. Paabo had been thinking about how to identify genes that had changed during human evolution to make speech possible, and FOXP2 seemed like a prime candidate. He and his coworkers sequenced the gene — that is, they figured out the order of the DNA bases that make up FOXP2 — in six different species. They found that it was one of the most stable genes they had ever studied; from mice to rhesus macaques to chimps, the protein produced by the gene is almost exactly identical, suggesting that the gene itself plays a fundamental role in animal function. But in humans the gene had undergone a slight modification. About 250,000 years ago, according to the scientists' calculations, two of the molecular units in the 715-unit DNA sequence of the gene abruptly changed. That's not long before modern humans first appeared in the fossil record.

Could the changes in FOXP2 have enabled modern humans to speak? And could articulate speech have given modern humans an edge over the Neanderthals and other archaic humans? That's certainly what some newspaper stories implied, labeling FOXP2 a "language gene." But Paabo and other scientists are more cautious. FOXP2 "is one of who knows how many genes that affect language ability," says Ken Weiss, an expert on evolution and genetics at Pennsylvania State University. The change in FOXP2 might have been entirely coincidental. Or the gene may be related to language indirectly — for example, by influencing coordination. And some scientists argue that language evolved much earlier than our version of FOXP2 and that archaic humans also had speech.

Still, Paabo's work on FOXP2 has raised fruitful questions. Researchers are genetically engineering mice with "broken" FOXP2 genes to see how disruptions in the gene might affect the animals. Also, researchers are splicing the human version of the gene into mice to see if it makes any difference. (So far, none of the mice have started talking.)

More recently, Paabo has taken an even broader view of the genetic changes responsible for our uniquely human traits. For example, mutations in individual genes like FOXP2 may not be the most important force in evolution. An even bigger factor may be changes in the genetic switches that turn on and off many genes at once. Paabo and his colleagues have been looking at the patterns of gene activity in humans, chimps, and other species. As might be expected, the brain has been a particularly active site for recent human evolution. Paabo's team finds that genes in the human brain have undergone more changes in how they are turned on than similar genes in chimp brains.

Paabo is also returning to one of his original obsessions. Using a fossil from a site in Croatia, he and his colleagues are trying to derive much longer Neanderthal DNA sequences — not just the DNA that runs the mitochondria, but the DNA that is responsible for building the rest of the body. Their goal is to reconstruct the entire genetic blueprint for making a Neanderthal. It's a technically daunting task, and Paabo estimates it will take about two years to finish. But being able to compare our genome with that of our evolutionary relatives could highlight key turning points in our evolution.

The ultimate goal of his research, Paabo says, is to identify the genetic changes that made us human. Of course, no historical event can ever be reconstructed completely. But by studying our DNA, scientists eventually will be able to say which genes changed, when they changed, and maybe even why they changed. At that point, we'll have something we've never had before: a scientifically plausible and relatively complete story of our biological origins.

About a mile north of the institute, down a dim alley and a flight of stairs, is a very old restaurant known as Auerbach's Cellar. In Johann Wolfgang von Goethe's 1808 epic play *Faust,* the Devil and Faust go drinking at Auerbach's. Shortly thereafter, Faust meets and talks with two apes — symbols, for Goethe, of human sinfulness and folly.

Faust sold his soul to the Devil for knowledge. Will the knowledge generated by studying our DNA place limits on the human soul? Will people come to see themselves as biological automatons bereft of compassion and morality? Will genetics "biologize" human relationships, so that we begin to define ourselves and others in terms of our DNA sequences?

Paabo worries about such possibilities. DNA studies have revealed how similar we are to other organisms, even such lowly creatures as worms and flies. These discoveries have emphasized the unity of life, Paabo says, but they also have been "a source of humility and a blow to the idea of human uniqueness."

Paabo, like most scientists, is an optimist. He believes that genetic knowledge will strengthen our commitments to each other, not rob us of purpose. Studying how we evolved may reveal why human beings suffer from diseases not found in other animals. He is particularly insistent that studies of the evolutionary origins of speech will help children who have congenital speech problems.

But Paabo also says that the possible benefits of research are not his principal motivation. "I'm driven by curiosity," he says, "by asking the questions where do we come from and what were the important events in our history that made us who we are. I'm driven by exactly the same thing that makes an archaeologist go to Africa to look for the bones of our ancestors."

MICHAEL PERRY

Health Secrets from the Morgue

FROM *Men's Health*

I'VE BEEN INVITED to watch someone pull the guts from a dead man. The man has died, as they say, before his time. He was in his forties. If you want to know why young men die young — and you'd like to avoid the same fate yourself — it makes sense to look over the shoulder of someone like Michael Stier, M.D. Dr. Stier is a forensic pathologist often called as an expert witness in court cases, and I am meeting him in a morgue near the University of Wisconsin medical school, where he is an assistant professor. In a short while, Dr. Stier will be up to his wrists in a fleshy cavity containing organs that only a few hours ago pulsed and squeezed and lent animation to a man's body.

It's been some fifteen years since I last observed an autopsy. I was in training as an emergency medical responder, and the morgue was located in a hospital in Eau Claire, Wisconsin. The man on the table was young. He had gone on a bender and died of exposure after wandering into the woods on a cold night. I carry only a few images: his shriveled gonads like misshapen gray clay, the medical examiner slicing the liver like a loaf of bread, and — in the only part of the procedure that struck me as creepy — the man's scalp being peeled up and rolled over his face.

Since then, I have observed and handled dead bodies of all sorts, from shooting victims to mangled loggers, so I'm not overly concerned about puking on Dr. Stier's plastic clogs. Still, it's one thing to observe mutilation while amped on adrenaline; it is quite another to watch as a body is methodically turned inside out and the organs removed, sorted, and filleted. You have to steel yourself a bit.

*

Generally, preventing your own untimely demise is simple enough (buckle up and slow down, for starters), but pathologists pick up on dead-man trends the rest of us never contemplate. They know you may not be doing all you can to avoid the horizontal refrigerator. They know that even if you are ripped or can run 5 miles, you may still be at risk of a fatal heart attack. They know that one of the most cunning killers of men hides out in the brain and can't be seen with a microscope.

I'm interested in the outcome of Dr. Stier's autopsy, but above all I hope to gather clues as to how I might postpone my own. Some of these clues will be in the corpse, but, as in any good detective story, the best of the rest I plan to gather from men who work the beat.

Andrew Baker, M.D., chief medical examiner for Hennepin County, Minnesota, an area that includes Minneapolis, gives me my first lead: when he pulls back the sheet, the face staring back at him is usually that of a man. "We see men of all ages with much greater frequency than we see women," he says. "That's taking all comers: gunshot wounds, drug overdoses, suicides, car crashes, and the occasional truly natural death of someone under forty."

What do we, as men in the full bloom of life, have most to fear?

Ourselves, apparently.

Dr. Stier's case — a man discovered dead in his garage — is still a lump waiting inside a black body bag when I emerge from the locker room in my scrubs. An assistant hands me a face shield and offers an odor-filtering mask. I detect a warm, sort of brown smell, right on the edge of foul. I once talked to a paramedic who entered a house around Thanksgiving and got hungry at the aroma of turkey — then went off it for years when he discovered an old man dead for days, with his forearm slow-roasting on a portable heater. I take the mask.

Dr. Stier arrives ten minutes later, fresh from a morning workout. "I've been working on my abs real hard," he says. Of medium height and stocky build, he explains that he recently dropped thirty-five pounds. Swimming, mostly. "See these?" he says, pulling a pair of XXL scrub pants from a pile. "They used to special-order them for me." Dr. Stier, thirty-eight, tends to lock his hazel eyes on you when he speaks. It is the gaze of a man used to looking at bad things without flinching.

Before examining the body, Dr. Stier stops to speak with a police detective who has photographs of the death scene. The dead man had been working on his pickup truck and appeared to have been asphyxiated by carbon monoxide. But the photographs reveal raw red patches on the man's torso. Uncertain about the source of these injuries, the detective has come along to view the autopsy. "We're wondering if he might have been struck or burned," he says.

When Dr. Stier's assistant unzips the body bag, the face that emerges is grossly swollen, and the tongue is protruding. Now and then a bloody bubble of escaping gas squeezes from the lips. Dr. Stier circles the body while speaking into a handheld recorder, describing what he sees. When I ask him what he's looking for, he says, "Everything." It's important not to zero in on the obvious and overlook something critical — hand injuries not consistent with mechanical work, for instance, or inconspicuous needle holes. He turns off the recorder long enough to tell the detective that the red marks are typical of the way skin loosens and "slips" during normal decomposition. He also points out that the dead man's skin is bright pink, a condition that can occur when hemoglobin — which normally carries oxygen in the blood — bonds instead with carbon monoxide. If the man had suddenly dropped dead — say, of cardiac arrest — he wouldn't have breathed in all that carbon monoxide.

The vehicle's gas tank was empty, and there were tools scattered about, but it's hard to imagine even a weekend mechanic working on a running car in a closed garage for more than a few minutes. There are other possibilities. Perhaps he had a stroke and was unable to move. "Or," says Dr. Stier, pointing to the photo of the scattered tools, "people sometimes disguise their own suicide scene to make it look like an accident."

It's those sudden "natural deaths" Dr. Baker mentioned that make a guy want to reach for his wrist and check his pulse. According to Randy Hanzlick, M.D., a professor at Emory University and the chief medical examiner of Fulton County, Georgia, the majority are tied to a man's ticker. "Usually there's some sort of unsuspected or premature cardiovascular disease," he says. "Or there's drug use — cocaine, for instance — that has affected the heart over the long term. And in a small number of cases, we don't find anything." The heart simply fails.

Dr. Baker's own hands-on examinations of cadaverous cardiac tissue have brought him to a similar conclusion. "The most common thing we see is just advanced coronary-artery disease that men didn't know they had," he says. "They're thirty-five years old, and their arteries look like they're sixty."

What can cause a man's heart to grow so old so early? Most often, a demon in his DNA. "You really have to be concerned about a genetic component when you have heart disease that young," says Dr. Baker, adding that when he sees early-onset atherosclerosis, he'll apprise the family of the danger that may be lurking in their genes. That danger could be familial hypercholesterolemia (FH), a disorder that impairs the body's ability to remove cholesterol from the blood. It's estimated that one in five hundred people has FH, though many cases go undiagnosed. Of the people in this group, significantly more young men than women will suffer what FH researchers describe as "premature cardiac death."

While an autopsy may be a lifesaver for the relatives of men felled by FH, it isn't exactly a frontline diagnostic tool. That's why it's recommended that all young men research their family's health history and have their cholesterol checked every five years, starting at age twenty. Any man who rings up an LDL cholesterol score of 200 milligrams per deciliter or higher — and isn't leading a sedentary, saturated-fat-filled existence — should ask his doctor for further tests, including genetic testing for mutations in the LDL receptor gene. Going home and getting naked is also in order. Men can uncover clues that they have FH by examining their elbows, knees, and buttocks for xanthomas, bumps that form when excess cholesterol piles up under the skin.

But if FH is difficult to diagnose, at least it's relatively straightforward to treat: a statin, that chemical antidote to high cholesterol, can help keep most afflicted men out of the morgue.

Dr. Stier's assistant picks up a scalpel and begins the autopsy by drawing an incision from the pubic bone to the sternum, where he bifurcates the incision, cutting toward each shoulder to form a Y. In the wake of the blade, skin and fat part with a delicate hiss and crackle. The assistant rolls the flesh back from the chest, then snips the ribs with a tool akin to pruning shears. The bones part with a wet crunch. When the last rib is cut, the assistant lifts the shield-shaped chest plate away to expose the thoracic space, then drapes open the belly skin to reveal the contents of the abdomen. Meanwhile, Dr. Stier is sucking fluid from one of the cadaver's eyeballs with a large-gauge needle. Because the eyes absorb drugs and alcohol at a different rate than blood does, the fluid can sometimes help determine the time of death.

The exposed innards are mostly all reds and yellows. It is bracing to be reminded of all the organs we carry packed within us, churning away no matter whether we are sleeping, eating, or driving screws into decking. The space management alone is fascinating — the way the heart nestles between the lungs, the way the lower lobes of the lungs curve above the diaphragm, the way the diaphragm curves above the liver. Dr. Stier draws the final blood sample and then straightens. "Now," he says, "we take everything out."

The organ-sorting portion of the autopsy is workmanlike, and fairly speedy. One by one, each is cut loose, removed, and weighed, with the results posted on a chalkboard at the foot of the table: SPLEEN 100 g, R KIDNEY 200 g, L KIDNEY 250 g, HEART 300 g, and so on. Before placing an organ on a steel tray for dissection, Dr. Stier examines it visually. "Sometimes you open a suicide victim and find advanced cancer," he says. "People get the diagnosis and end their lives because they can't face the treatment or don't want to deal with the course of the disease."

Some of the organs — the lungs and liver, for instance — hold their shape well, while others, like the stomach (which Dr. Stier empties and checks for pills; there are none), look like so much uncooked meat. Because he was a lifelong smoker, the man's lymph nodes are stained brown, and his lungs are stippled with anthracotic pigment — clotty black rivulets that lend the lungs the speckled look of a rainbow trout. Three feet down the table, Dr. Stier's assistant is "running" the small intestine: stripping it out of the abdomen, checking every inch for abnormalities, and then spooling it into a plastic bucket.

Dr. Stier holds the dead man's heart in one hand; in the other he holds a small knife, with which he slices crosswise through the left anterior descending artery every 2 or 3 millimeters, creating a series of small serrations. "When a man dies prematurely of a heart attack, this is the artery most likely to be obstructed," he says. "We call it the 'widow-maker.'" It looks nothing like the red rubber tube so fulsomely rendered in American Heart Association brochures. After each cut, Dr. Stier rolls the blade over as if he were laying out a slice of cheese. Section after section, the artery is clear of the pasty white-yellow gloop you'd expect from a man who died while carrying a pack of cigarettes. "A drinker without cirrhosis is more common than a smoker without atherosclerosis," says Dr. Stier, plainly delighted to point out an exception to the rule.

Sometimes the crushing pain in a man's chest isn't a heart attack but a pulmonary embolism, in which a blood clot travels from one part of the body (usually the lower leg) to block an artery in the lung. But while the source of the symptom may be different, many times the end result is the same: an early exit.

"The youngest I've seen was seventeen," says Dr. Stier. "Usually there were risk factors, including extended periods of immobility — such as an airplane flight — or an injury like a severe sprain or fracture that required immobilization in a splint or cast."

Even though the flying-is-dying connection to pulmonary embolisms is the best known — their frequency prompted doctors to coin the term "economy-class syndrome" — some researchers believe that a desk job can be just as dangerous. Last year scientists in New Zealand reported on four cases in which young men developed either a pulmonary embolism or deep-vein thrombosis — the initial formation of a clot — after remaining seated at a computer for between three and six hours at a clip. The researchers even came up with their own catchy name for the phenomenon: "seated immobility thromboembolism," their contortion to arrive at the acronym *SIT*.

The only good news about this white-collar man-killer is how simple it is to prevent. If you tend to get glued to your CPU, take regular walking breaks. If you spend your life at 35,000 feet, drink water (dehydration increases the risk) and walk the aisle every two hours. And if you're seatbelt bound? Press your heels against the cabin floor several times an hour to help push blood through your calves. As a bonus, a recent study in the *International Journal of Cardiology* shows that such isometric exercises can reduce high blood pressure — another invisible menace that attacks men much earlier than it strikes women and can put us on the fast track to a formaldehyde injection.

Dr. Stier asks me to move in closer as he prepares to cut up the dead man's liver, which is mottled with yellowish tan patches. "The liver should be reddish brown," he says while running his gloved finger over the surface. "See how it's greasy? Almost oily? These are signs of alcohol damage. If you stop drinking at this stage, the liver will heal, but once scar tissue accumulates, the changes are permanent." Dr. Stier's knife moves through the organ smoothly — in more advanced liver disease, it can be physically difficult to push the blade. Long-term alcohol abuse can also ruin the pancreas, causing it to shrink and harden until it has a chalky feel. Dr. Stier checks, and the man's pancreas appears normal.

With most of the chest and abdominal cavity empty, the deflated tube of the descending aorta is visible. Dr. Stier slits it lengthwise. The aorta falls open to reveal smooth, curdlike protuberances clinging to the wall like clumps of egg white. These are atherosclerotic plaques, and seeing them up close, I wonder how many I'm growing. These are in no danger of blocking the aorta — it's roughly the diameter of a garden hose — but you can see how they would be deadly in a cardiac artery.

I notice that Dr. Stier's assistant has placed one hand on the crown of the dead man's cranium. In the other hand he holds a scalpel, and as he runs it ear-to-ear across the back of the man's head, I prepare for the sound of the bone saw.

If you spend enough time stuffing people into the back end of an ambulance, as I have, it's bound to alter your behavior. I love motorcycles, but you'll never catch me astride one — at the last crack-up I responded to, I had to pull slivers of femur out of a car door.

Having seen people killed as they were getting the mail, I also wear my bicycle helmet for the one-block ride to the post office. If I take the car three blocks to the gas station, I wear a seatbelt. My experience with the local fire department has left me with a pathological aversion to electric space heaters. On issues of personal safety, I am, frankly, a fussbudget. And yet, on the personal health front — Exhibit A: late-night gas station doughnut binges — I constantly lapse into bad habits. Perhaps if I spent five days a week taking dead people apart to see why they died, I would take better care of my own insides.

"I try to eat as healthy as I can," says Dr. Baker. "I run, bike, and swim, and obviously I would never take up cigarette smoking or drink excessively. The average person would think smoking is a bad idea, but for a forensic pathologist, that point is reinforced day after day." The doctor also makes a point of seeing a doctor. "I see my own internist," says Dr. Baker. "I want someone who takes care of living people counseling me about my weight, my exercise, and my cholesterol."

Dr. Hanzlick's perspective is even more intriguing. He admits that his job has little impact on his diet and personal behavior, but it has made him less comfortable about heights. "I think it has to do with seeing people over the years who have fallen off high structures. Oh, and lightning bothers me now, too." His fears are not unfounded: according to the National Oceanic and Atmospheric Administration and the Centers for Disease Control and Prevention, 84 percent of lightning fatalities are male — most between twenty and forty-four years of age.

It's relatively easy to get at the organs in your abdomen and thorax. Your brain, on the other hand, is in a lockbox. Dr. Stier's assistant buzzes around the circumference of the cranium with a bone saw and tips the skullcap back as if it's on a hinge. Dr. Stier puts a hand on each side of the cerebrum and lifts it like custard from a Jell-O mold. The brain is beginning to decompose and comes apart in Dr. Stier's hands as he searches for any evidence of infection, cancer, trauma, or bleeding. He finds nothing abnormal. The brain is weighed and dissected, and the skullcap and scalp are replaced.

Of all the organs that can kill a young man, the brain ranks high. For one thing, it is responsible for decisions like "Hey! Let's bumper-surf naked!" But Dr. Baker warns that the male brain often incubates another deadly danger. "Untreated depression is a significant part of many of the deaths we see. Just as I would not want young men to blow off chest pain, I'd hate for them not to get their depression treated, whether because of finances or social stigma."

The stats bear him out: men are four times more likely to die as a result of suicide than women are. Some guys, of course, are more at risk than others. In a 2005 study from Johns Hopkins, researchers discovered that even though male physicians tend to adopt healthy behaviors that give them mortality rates 56 percent lower than the rest of us, they're much more likely to write out their own death certificates.

"I think it's stress-related," says Dr. Stier. "Doctors are overachievers surrounded by overachievers. I see it in my med students — if they have problems, they try to mask them. That mindset continues into practice." Mix in the male propensity for suicide and an educated understanding of the most effective methods of death, and you have the recipe for some serious malpractice.

If this were *CSI: Wisconsin,* Dr. Stier would have his *Aha!* moment about now. As it is, we are left to speculate. All signs point to carbon monoxide poisoning (the toxicology results are necessary to make a definitive call), but was this man a victim of bad luck or careful planning? "I determine cause of death," says Dr. Stier, a pathologist. "The coroner establishes the manner of death." Armed with Dr. Stier's report, the coroner will interview the deceased's family and friends in a search for clues. Intentionally or accidentally, the man is dead. The assistant tucks the organs inside the human rind, stitches up the Y, and rolls the body from the room.

When I walk out of the morgue and into the light of a bright fall morning, it's good to see the sun. The signs of wear I saw in that man — the mottled liver, the clumps in his aorta — hadn't killed him, but they were hard evidence of what we accumulate over time and through denial. Above all, that paradoxically healthy cardiac artery niggled at me. Do my arteries look like that? Or are they crammed with sludge? I have never smoked, but my last two cholesterol tests were high and higher.

First thing tomorrow morning, I'll report to my general practitioner for a thorough physical. Once you've seen someone's guts in a bucket, it's difficult to think of your skin as anything but a bag full of trouble.

HEATHER PRINGLE

Hitler's Willing Archaeologists

FROM *Archaeology*

In 1935, Heinrich Himmler, head of the Gestapo and the SS, founded an elite Nazi research institute called the Ahnenerbe. Its name came from a rather obscure German word, Ahnenerbe *(pronounced AH-nen-AIR-buh), meaning "something inherited from the forefathers." The official mission of the Ahnenerbe was to unearth new evidence of the accomplishments and deeds of Germanic ancestors "using exact scientific methods."*

In reality, the Ahnenerbe was in the business of mythmaking. Its prominent researchers devoted themselves to distorting the truth and churning out carefully tailored evidence to support the ideas of Adolf Hitler, who believed that only the Aryans — a fictional "Nordic" race of tall, flaxen-haired men and women from northern Europe — possessed the genius needed to create civilization. Most modern Germans, he claimed, were descended from these ancient Aryans. But scholars had failed to uncover any proof of a such a master race lighting the torch of civilization and giving birth to all the refinements of human culture. The answer to this problem, in Himmler's mind, lay in more German scholarship — scholarship of the right political stripe. So he created the Ahnenerbe, which he conceived of as a research organization brimming with brilliant mavericks and brainy young upstarts who would publicly unveil a new portrait of the ancient world, one in which Aryans would be seen coining civilization and bringing light to inferior races, just as Hitler claimed.

Before 1938, the Ahnenerbe largely confined its studies to ancient texts, rock engravings, and folklore. But in February of that year Himmler transferred the Excavations Department of the SS into the Ahnenerbe. Himmler had created the department to sponsor or direct archaeological digs at major sites in Germany. He intended these excavations to be exemplars of German research where SS men could be trained in the science of recovering the ancient Germanic past from the ground. The department had financed eighteen excavations, from an ancient hill fortress at Alt-Christburg in Prussia to a major Viking trading post at Haithabu in northern Germany, not far from the Danish border. The Excavations Department brought new scientific acumen to the Ahnenerbe. Its staff consisted of dirt archaeologists trained in analyzing bits of ancient stone, bone, and ceramics. This new expertise, Himmler hoped, would help reconstruct the lives of Germany's ancestors before the first written histories and greatly extend knowledge of the mythical "Nordic" race. Indeed, one of the Ahnenerbe's most ambitious young researchers, archaeologist Assien Bohmers, claimed he could trace "Nordic" origins all the way back to the Paleolithic era in Germany, when woolly mammoths and cave bears wandered the chill tundra.

BOHMERS WAS a Dutch national with a remarkable talent for dissembling, scheming, and self-advancement. Having studied geology, petrology, and paleontology at the University of Amsterdam, he was an expert on soil and sediment analysis and on interpreting the complex stratigraphy of caves. He also took a strong interest in racial research, conducting what he called "some pretty detailed studies" on the "Nordic" race. He traveled widely, doing geological and archaeological fieldwork in southern France and northern Spain, Sweden, and Norway.

After graduation, the twenty-six-year-old researcher had his pick of jobs in the European oil companies. But in April 1937 he decided to apply for a post as a Paleolithic archaeologist in the SS Excavations Department. Bohmers sent off his curriculum vitae, as well as a small head-and-shoulders photograph. The latter, clearly intended to demonstrate his racial suitability, showed a young man with a thatch of blond hair, deep-set eyes, and the good looks of someone who grew up with plenty of fresh air and wholesome food.

His application arrived at a rather opportune time in the SS headquarters in Berlin. A German scholar, R. R. Schmidt, had just begun excavations at Mauern in Bavaria. The site consisted of a network of five caverns, and near the eastern entrance of the middle cave, Schmidt had found something very important — a bed of soil stained red with ocher, a mineral pigment. Spread out along this red layer was the skeleton of a woolly mammoth blanketed with ivory beads and beautifully fashioned stone weapons and tools.

The ivory beads and the finely worked stone tools were the product of the Cro-Magnons, early *Homo sapiens sapiens* who hunted reindeer and other big game beginning some 35,000 years ago. The Cro-Magnons had often chosen red ocher as a pigment for their cave paintings and for the ceremonies they performed for the dead. In all likelihood, the mammoth skeleton at Mauern had been part of some Cro-Magnon religious rite. This news seems to have excited local Nazis greatly, for one of Germany's most prominent racial researchers, Hans F. K. Günther, had declared that the "Nordic" race arose from the earlier "Cro-Magnon race," based on physical similarities he somehow discerned in the two groups.

In August 1937 the SS Evacuations Department took control of the dig at Mauern. The head of the department, Rolf Höhne, became the site inspector, and a unit of the Austrian SS — an underground organization at the time — began preparing the area for further excavations. By September, however, Höhne realized that he needed an experienced Paleolithic excavator, someone capable of retrieving every possible scrap of evidence concerning the "Cro-Magnon race."

In the fall of 1937, Bohmers arrived at Mauern to run the excavations. He studied the lay of the land, examined the cave floors, and scrutinized the exposed stratigraphy. As the days grew shorter and cooler, the team dug deeper into the sediments near the entrance to the middle cave and struck more fossilized bone and ocher. It was the skull and several vertebrae and ribs of another woolly mammoth.

Someone had broken off the mammoth's long ivory tusks and scattered precious caved-ivory pendants and a collection of drilled fox, wolf, bear, and reindeer teeth — all ready to be hung on a necklace — near the mammoth bones. Then he or she had scattered red ocher nearby and lit a fire, for the ground was littered with bits of charcoal. Bohmers was fascinated. He concluded that the area was a cult site.

As the excavators dug further, they encountered ancient hearths and the refuse of a different group of visitors: the Neanderthals. In these layers the team found hand axes, scrapers for cleaning animal skins, and spearheads suitable for handheld spears wielded like bayonets. There was nothing terribly unusual or striking in any of this — except for one curious find. In the upper portion of these layers, Bohmers and his team unearthed thirty-three leaf-shaped points, tapered at both ends. These points resembled the streamlined spearhead of a technologically advanced weapon: the throwing spear or javelin favored by much later Cro-Magnon clans. Armed with this weapon, a human hunter could slaughter dangerous prey — animal or human — from afar rather than closing in and risking serious injury. According to his calculations, which later proved to be quite wrong, the curious tapered leaf points dated to more than 70,000 years ago, to a period between the last two glacial ages.

The young researcher chose to interpret this finding in a very dramatic way that would appeal to the SS. Since he believed inherently in the superiority of the "Cro-Magnon race," he assumed that only Cro-Magnon hunters were intelligent enough to invent leaf points and throwing spears. Therefore the presence of these projectile points in the stratified layers at Mauern proved that Cro-Magnon people stopped there more than 70,000 years ago. By this reasoning, Mauern was the earliest Cro-Magnon site in the world — by tens of thousands of years. It was a very bold interpretation for a young prehistorian, for the business of dating cave floors before the advent of radiocarbon dating was a particularly difficult one.

Bohmers proudly announced his results in a handwritten letter to the Ahnenerbe, which had just absorbed the SS Excavations Department. "Your assignment," he declared in July 1938, "has brought me face to face with a conclusion that is extremely important for all future Stone Age research and especially for all areas of study which are concerned with the predecessors of the Germanic tribes — the so-called Cro-Magnon race. Until now almost every German, and without exception every foreign investigator, has assumed that the race migrated to Europe from somewhere in the East. The excavations at Mauern . . . have revealed for the first time the key that proves that the Cro-Magnon race must have developed in greater Germany." This news, he knew, would please Himmler mightily, for it neatly eliminated a potential source of foreign blood in the pedigree of Germanic ancestors. Moreover, it seemed to confirm another favorite boast of SS leaders: that all important advances took place first in northern Europe.

When word of the young prehistorian's discoveries at Mauern

reached Himmler in late July 1938, he was fascinated. Although he was mired in work, engineering a major new anti-Jewish campaign in Germany and drawing up plans for covertly undermining the government of Czechoslovakia, he asked to read the young archaeologist's papers even before they were published. Bohmers was well aware, however, that the European scientific community would be considerably less enthralled by these ideas. Indeed, his foreign colleagues would need serious convincing, for Bohmers knew that many European scholars scoffed at the science of the Third Reich. "Everywhere outside of Germany," he noted in a letter, "German science is seen as chauvinistic, without foundation." To deflect such criticism, Bohmers wanted to talk with his foreign colleagues and examine their collections of early Cro-Magnon tools. He also wanted to tour the painted caves in France — one of the best surviving records of the Cro-Magnons and their religious rituals, a subject that intrigued him.

Bohemers ventured first to Brussels to examine a large collection of flint tools and pebbles gathered by Belgian archaeologist Aimé Rutot. He then moved on to Paris and the Institute of Human Paleontology, where he spent two weeks sifting through the institute's collection of flint tools. In the hallways and offices, he introduced himself to the institute's renowned staff and paid his respects to a small, round-faced priest in a worn black cassock who always made time for young researchers. Abbé Henri Breuil was the world's foremost expert on Paleolithic cave art.

In 1901, Breuil and two companions had discovered ancient cave paintings and engravings in the Dordogne region of southern France, and it was Breuil who helped convince European prehistorians that the cantering horses, charging bison, leaping stags, and lumbering cave bears were indeed the work of Paleolithic artists. The priest had just returned from a month in the Pyrenees, copying an assortment of painted figures at Les Trois-Frères, one of the most important French caves, work he dearly hoped to finish before war brought cave-art research grinding to a halt.

Despite his worries, however, Breuil took time to see Bohmers. Almost certainly the genial priest rolled and lit a crumbling cigarette — he was seldom to be seen without one — as Bohmers described the leaf points, red ocher, ivory beads, and mammoth bones of Mauern. In all likelihood, Bohmers soft-pedaled his racial interpretations. Then Breuil, a generous-spirited man who could not possibly have guessed the extent of Bohmers's Nazi ambitions, helped to arrange a rare honor for the young Dutch researcher — a visit to Les Trois-Frères. The ancient site lay on a private estate, closed to the public. The owners permitted just sixteen or so people a year — archaeologists for the most part — inside.

Traveling south by train, Bohmers first visited the painted caves of the Dordogne. Then he pressed on to the rugged Pyrenees and the small French village of Montesquieu-Avantès, where the mysterious Les Trois-Frères awaited him. The cave had haunted Breuil for years. The abbé believed that it had long served as a subterranean shrine, a place where generations of sorcerers and enchanters came to pray to their gods and perform their ancient magic. As a spiritual man, Breuil felt certain that only initiates had been permitted to venture into its dangerous subterranean passages and enter the chamber he called the Sanctuary.

Even after a decade of meticulous study, Breuil still marveled at this innermost chamber. Along its walls, human hands had incised and painted a strange phantasm of creatures. In the darkness, a swirling maelstrom of superimposed bison, horses, ibex, rhinoceros, and a host of mythical or masked creatures, half human, half animal, capered and pranced.

The Sanctuary teemed with life and vital energy, and above all this tumult, beyond all the charging, rearing, and swaying, someone had engraved and painted a tall antlered man with a stag's ears and a horse's or wolf's tail. Breuil dubbed this surreal figure "The Sorcerer." He theorized it was a primeval divinity of the Paleolithic world — the spirit that controlled the fertility of game and the success of the hunt.

As the light of his lanterns flickered across the walls of Les Trois-Frères, with their strange unearthly visions, Bohmers gazed at the tumult of engravings. He had arrived at the center of a mystery, at a place where Himmler and where so many other Nazis had long dreamed of standing — in the shrine of the ancient dead, in the dark embrace of their supposed "Nordic" ancestors.

That winter Bohmers basked in the warmth of Himmler's favor. The SS leader lapped up his theories, so much so, in fact, that Himmler asked him to report on his findings at a conference of the

SS-Gruppenführer, the major-generals who would faithfully carry out orders for the Final Solution. Himmler also found time to take Bohmers aside at a gathering to convey his personal views on the subject of human evolution. It must have been an instructive conversation. As Bohmers later reported, Himmler dismissed outright the notion that the human race was closely related to primates. He was also outraged by an idea proposed by another German researcher that the Cro-Magnons arose from the Neanderthals. To Himmler, both these hypotheses were "scientifically totally false." They were also "quite insulting to humans."

Bohmers must have felt stunned. He too believed that the Cro-Magnons had evolved from the burly Neanderthals, based on the finds from Mauern, and he was unaccustomed to taking scientific direction from amateurs. He had clearly reached a major turning point in his career. He could ignore Himmler's notions and pay the inevitable price for his stubbornness: scholarly obscurity. Or he could quietly drop all talk of evolution and build an important career in Nazi Germany with the help of the Ahnenerbe. He chose the latter. He wrote in a letter to the institute's managing director on March 12, 1939, that he "always agreed with the Reichsführer's and the Ahnenerbe's position" on matters of human evolution.

Bohmers planned to create a satellite Ahnenerbe research center for Paleolithic studies in Munich. He believed himself to be the natural choice for its head, and he confidently predicted that the new center would soon outshine the Institute of Human Paleontology in Paris and "become the leading research site in the world." Indeed, his plans for the future even encompassed something that would have astonished Abbé Breuil and his colleagues in Paris. Bohmers, it seems, dreamed of taking a leading role in French cave-art research. The ancient murals and reliefs, he noted in an official 1939 memorandum, could shed much light on the subjects dear to the hearts of the Ahnenerbe staff, such as "symbol studies," the "history of hunting methods, i.e., wild horse, bear and mammoth hunting," and "knowledge about the magical customs of the Cro-Magnon race, i.e., hunting magic, initiation rites, etc." The French, in his view, had shamefully neglected this priceless legacy. He thought their scientific publications on the caves scanty and inadequate and their preservation methods a disgrace. He noted that "the named caves are owned by farmers, or are managed by farmers, who don't have the faintest idea about the cultural value of the rock art." As a result, American and French tourists had swarmed in droves through the caverns and run amuck, scratching the walls with penknives.

To Bohmers, the solution was perfectly obvious. If the legacy of the ancestral "Nordic" race was to be protected and preserved for future generations in Europe, then his new institute would have to take the lead in cave-art research. "The Ahnenerbe," wrote Bohmers, "would serve the whole culture well if it were to succeed in preserving these valuable documents from their eventual destruction by photographing, drawing or creating casts for the future."

Bohmers, it seems, already dreamed of a day when the swastika would fly over France.

German aggression in Europe did indeed create important new opportunities for Bohmers, although not the ones he had schemed for. In October 1938 the Third Reich annexed the Sudetenland in Czechoslovakia, and ten months later Bohmers took over the excavation of one of its most important Paleolithic sites — Dolni Vestonice. A prominent Czech archaeologist, Karl Absolon, had dug in the famous site since the 1920s, unearthing a trove of Ice Age figurines that included small ceramic cave lions and cave bears. But Absolon had recently retired, and in his place the Ahnenerbe's senior management appointed Bohmers. It was a great plum for a young Dutch archaeologist, but Bohmers's three field seasons at Dolni Vestonice between 1939 and 1942 proved disappointing: his team uncovered few artworks to rival Absolon's spectacular finds.

By early 1943, however, the attention of the Ahnenerbe was focused elsewhere — on a series of war crimes ordered by Himmler. Some of the institute's staff plundered the rich archaeological museums of the Soviet Union, adding to Himmler's private collection of artifacts. Other researchers directly supervised medical experiments in the concentration camps and arranged for the slaughter of Jewish prisoners. There was also serious talk in the Ahnenerbe offices of having Karl Absolon "liquidated," for he had become a voice critic of the Nazi regime in Czechoslovakia. But the senior staff feared the outrage of the international scientific community and thus never approved orders.

By the summer of 1943, Bohmers had largely abandoned the work at Dolni Vestonice. He spent most of his time traveling and dabbling in Nazi politics in occupied Holland. He aspired to become a senior Nazi official along the northeastern coast, and at one point he threatened to murder a Dutch excavator who refused to become involved in his political schemes.

The end of the war brought Bohmers crashing to earth. A Dutch colleague reported him to the Allied forces, noting that Bohmers had taken personal possession of a large collection of important artifacts. The archaeologist spent nine months in internment, ultimately convincing his captors that he was guilty of no crimes. He was released in February 1946.

Surprisingly, Bohmers still had influential contacts in Holland, a country that despised Nazi collaborators. In 1947 the renowned Biological-Archaeological Institute at the University of Groningen hired him as a researcher. There he went about his work excavating Dutch sites with professionalism, but his personal behavior grew increasingly furtive and erratic. During the saber-rattling of the cold war, he became obsessed with the idea that the Soviet Union would invade the Netherlands. He was particularly worried about being captured. He sold his house, bought a large seaworthy yacht so he could flee to Scandinavia in the event of a Soviet invasion, and stocked the yacht with firearms, all purchased illegally from an assistant who was stealing artifacts from the institute. After a police investigation, the University of Groningen suspended Bohmers, and in January 1965 he asked to be dismissed. For a time he marketed the crafts of traditional Dutch wood-carvers, then moved to Sweden. He died there in 1988.

Like many other members of the Ahnenerbe's staff, Bohmers attempted to reinvent himself in his later years as a fierce opponent of the Nazi regime. In 1972, for example, he gave a talk to a group of Dutch university students about his years in the Ahnenerbe. By that time he had recast the entire history of the research organization. "Dr. Bohmers gave the story that he had never been a Nazi," one of the audience members later recalled. "He said he was a member of a resistance group of highly qualified scholars who formed a kind of secret society within the Ahnenerbe opposed to National Socialism. They were against Hitler."

JONATHAN RAUCH

Sex, Lies, and Video Games

FROM *The Atlantic Monthly*

MICHAEL MATEAS IS the sort of person who once built an artificially intelligent(ish) robot houseplant that monitored your e-mail and changed shape to reflect the mood of what it read — if that sort of person can be said to be a sort. This was in 1998, when Mateas was a doctoral student with some avant-garde ideas. *Office Plant #1,* as the creation was called, grew and shrank and blossomed and hibernated and waved its piano-wire fronds as it "fed" off e-mail traffic. Naturally, it also whistled, sang, moaned, and complained. Not long after building *Office Plant #1,* however, Mateas set it aside. He became interested in bigger things, like creating a new art form.

Meanwhile, Andrew Stern, a programmer and designer at a now-defunct video-game studio, was building artificially intelligent(ish) virtual pets. They were called Petz, and for a while they were a hit in the video-game industry. First came Dogz, in 1995, then Catz, and eventually Babyz, all adorable animated creatures that lived on your computer's hard drive. As Stern worked on making the virtual creatures emotionally appealing and realistic to play with, he began giving them artificial minds: goals, personalities, memories. It dawned on him that he wanted to work with adult characters in life-like relationships. He became interested in bigger things, like creating a new art form.

Not long after Petz debuted, Stern began attending some of the same conferences on artificial intelligence that Mateas haunted. It was probably inevitable that Stern, presenting his intelligent(ish) virtual pets, would run into Mateas, presenting his intelligent(ish) robot plant. It didn't take long for them to recognize each other as kindred spirits.

In certain rarefied circles of AI academia and video-game design, people sometimes theorize about a computer program that would combine the graphical realism of a modern video game with the emotional impact of great art. "Interactive drama," the concept is called. It might contain artificial people you could converse with, get to know, and love or hate. It might engineer dramatic situations, complete with revelations and reversals. Entering this world, you would feel as if you had been thrust into the midst of a soap opera or a reality TV show.

"I had some idea how to do it," Stern says. Mateas, for his part, had dreamed since childhood of building artificial humans. It occurred to him that he could advance his dream by building artificial *actors*. What better way to teach a computer to act human, after all, than by teaching it to act?

In 1998, emerging from a hot tub at a conference in Snowbird, Utah, Mateas and Stern decided to collaborate. "As Andrew and I talked," Mateas recalls, "we sort of egged each other on to jump as far out of the mainstream as possible." They resolved to create a game that would put a *not* in front of every convention of today's video-game industry. They looked upon their game as a research project and figured that building it would take two years. It took more than five. Now they are starting on a larger version, this time a commercial game.

They think interactive drama has the potential to be to this century what cinema was to the last. When I spent a couple of days getting to know them recently, I asked why they're not trying something more modest, such as making the characters in today's video games more lifelike. "That's a sort of incremental innovation that I think neither of us is interested in," Mateas replied. "We're interested in revolutionary innovation."

If today's video-game industry were a person, it would be at what people used to call "that awkward age." Suddenly, like a teenager with long legs and short pants, it finds itself grossing $31 billion this year in revenues worldwide, according to the business consultancy PricewaterhouseCoopers, and nearly $10 billion in the United States alone. If the industry keeps up its growth, Pricewaterhouse expects it to rival the global recorded-music business by about 2010. Yet the video-game industry, for all its swagger and success, remains something of a niche player. In the United States, it is smaller than the theme park and amusement park industry; according to Pricewaterhouse, its rapid growth would still leave it, in 2010, about a third the size of the film, radio, or book industry, and about a seventh the size of the television industry.

A lot of people play games now, and not just kids: the average gamer, according to the Entertainment Software Association, is thirty-three years old. But while just about everyone regularly listens to music or reads books or watches movies, many adults never pick up a joystick. Only about a seventh of game titles sold in 2005 were the racy or violent stuff that draws an *M* (for "mature") rating; the stereotype that video games are nothing but antisocial savagery is just that — a stereotype. Puzzles, pets, strategy games, and social games abound. But it's true that the adrenaline-pumping, youth-oriented genres dominate. According to the ESA, almost three quarters of the best-selling games on the market are in the fighting, shooting, racing, action, and sports genres. A sexist commentator might call it boy stuff.

The graphics of the best modern games are stunning, and their "physics" — their power to create a world that feels real as you move about in it — hardly less so. But the industry is rife with game designers who complain of "sequelitis" and creative underachievement. "Will we address an excruciatingly audience-limiting lack of diversity in our content?" wondered Warren Spector, one of the industry's leading developers, in a recent article in *The Escapist,* a video-game magazine. "I can see us limiting ourselves to the same subset of adolescent male players we've always reached. And if we do that, it's back to the margins for us."

"There's no drama genre, there's no comedy genre," Andrew Stern told me recently. "What exists right now are action movies, basically." He might have added, *Silent* action movies. The video-game industry's annual trade show in Los Angeles, called the Electronic Entertainment Expo, or E³ for short, is one of the loudest places I have ever been. Also one of the most silent.

This year's show occupied all of L.A.'s cavernous convention center. Its thousands of microprocessors and liquid-crystal displays and sound systems burned enough electricity to power a good-sized suburb. Take the crowds of Times Square, add the high-tech dazzle of Tokyo and the floor-shaking decibels of surround-sound cinema, throw in Vegas-style showgirls (known in the trade as "booth babes"), and you have some idea of E^3.

Drifting through the show last May, I saw many shallow games and many derivative games: superheroes dueling with giant robots, skateboarders flashing Nike logos, boxers throwing punches amid showers of sweat and spittle, warriors trudging through jungles and snowscapes. Joining one particularly long line, I found myself in a small, darkened room where a designer was debuting Midway's John Woo Presents: Stranglehold. Fighters were demolishing everything in sight. "Look at the state of the teahouse, just massive destruction," said the designer lovingly. "And it never looks the same twice." As he emphasized how realistically each bullet splintered the walls, a male connoisseur in the audience called out, "Aim for the head!" (The audience in this demo, and at the show generally, was at least 80 percent male.) Even the schlock, however, exhibited striking craft and ingenuity, and I came across some astonishingly imaginative games, including an alien-invasion shooter (Capcom's Lost Planet: Extreme Condition) whose visuals were so compelling that I was helpless to tear myself away.

It was only after I left the hall that I realized there was something odd about all the noise. The thunderous sound effects were masking the absence of conversation. In real life, much of what's interesting involves talking to people. The characters in games could deliver scripted lines like "I'm ready to kick some ass!" or drop prerecorded comments on the action, but conversing with me or each other was completely beyond them. It occurred to me that if video games seem inhuman, that is because they lack humans. Their esoteric syntax is an artifact of a stunted environment in which blasting someone's head off is easy but talking to him is impossible.

A month later, I asked Andrew Stern what he thinks of E^3. "I shake my head a little," he replied. "All this effort and money being poured into all this derivative and uninspired work. I'm bored and slightly disgusted." Few in the mainstream industry would express disgust with their product, but many designers, being intelligent and creative people, feel that they have made much less of their powerful medium than it could be. They are vexed by a sense of underachievement. As Will Wright, the most famous and successful

American game designer, told a crowded session at E^3, "Interactive design is a really large box, and we've really only explored one little tiny corner of that box." David Cage, another prominent designer, told another audience, "What strikes me in this industry is, there's just a real lack of meaning in general."

Meaning is the catalyst that turns action to drama. Meaning requires words, not just sounds. It requires characters, not just figures. It requires dramatic shape: a sense that the action is leading to some transformation or resolution. It is what Stern and Mateas resolved they would bring to video games.

Michael Mateas is an assistant professor in the computer science department of the University of California at Santa Cruz, where his duties include launching a new undergraduate degree program in games. He wears two earrings and keeps his bushy brown hair tied back in a long ponytail. His body is small and his head is large, so from a distance one could almost mistake him for a boy. His pale green eyes are piercingly intense, though their intensity is leavened by his beaming smile. He thinks of himself as equal parts artist and computer scientist, and he manages to look both roles.

Stern, by contrast, is so average-looking that he is hard to describe: medium height, thinning brown hair, soft features, an introvert's undemonstrative manner. He could vanish into any American crowd. Nonetheless, as the three of us talked, it was Stern who emerged as the dominant personality, partly because he has an artist's fierce sense of aesthetic rectitude. Economy, elegance, formal coherence: these are personal matters to him.

Stern is thirty-six and lives in Portland, Oregon. He grew up in various cities along the East Coast. Mateas is forty and grew up in Carson City, Nevada. In some respects, their childhoods ran in parallel. Both discovered video games as children, in the 1970s, when the very first games appeared. They haunted the arcades in the malls; they pounced on the Atari 2600 console when it appeared, in 1977. Not content with playing games, they soon began programming them. At fourteen, Mateas wrote an adventure game. Stern, whose brother kept a pet rabbit named Bonny, made a game called Bonny Attack, in which the player flew the rabbit around and dropped turds and urine on jumping cats. In a high school essay, Mateas announced his intention to become a big-time designer building games on a "new kind of digital logic circuit based on three-valued logic." Stern, meanwhile, was getting interested in film and computer animation. Using a Handycam, he began making movies that blended live action with animation.

Then, in college, they both lost touch with video games. "I would still play them," Stern says, "but they started to feel a little juvenile. I was getting into filmmaking, stories with real characters, adult characters, about psychology and emotion, and games weren't addressing those things. Once you get into your late teens or early twenties, you realize there's a lot more out there in terms of art and literature and you lose interest in action-oriented entertainment."

Mateas decided he would be a scientist, pursuing his longtime dream of artificial intelligence, and he went for his doctorate. Stern was rejected by film schools and wound up taking a job at a game studio. His work on the Petz games kindled his interest in artificial intelligence, the essential ingredient of believable characters, whether animal or human. Mateas's work on artificial intelligence, meanwhile, had rekindled his interest in games. The AI dream was about building believable virtual people, and games seemed the ideal stage to test them on.

By the time their paths crossed, their thinking had already converged. They soon began plotting their anti-game. Instead of making a game about action figures in elaborate but childish game worlds, they would make a story about adult characters and adult relationships. Instead of firing bullets at the characters, the player would fire words. The player would *talk* to the characters — in ordinary English, input with a keyboard rather than a joystick. And the characters would talk back, to each other and to the player. This meant — and they gulped to think of it — that their game would need to speak and understand natural language. That, in itself, is one of the great challenges in AI. But they didn't intend to stop there.

Conventional games create vast, immersive physical environments. The new game would all take place in a single indoor space, like a black-box theater stage. Instead of taking fifty hours to play, their game would take twenty minutes. Instead of advancing through levels without telling a story, the game would provide a compact, complete dramatic experience, like a one-act play. "We envisioned something where you could come home from work and play it from beginning to end, just like you come home from work and watch a half-hour television show," said Mateas. "You could come home and have a half-hour interactive drama experience. It's complete in itself, it takes you on an arc. It entertains. But then the next day you could come home from work and play it again and make something different happen." Instead of offering the player menus of quests or options, their game would seem to flow as naturally as life.

When Mateas, still a graduate student, told his adviser what they intended, the adviser replied that such a game would take a team of ten people ten years to build. The technology didn't exist. Commercial game design often employs teams of dozens, and here were two guys, one a grad student and the other self-employed (Stern eventually quit his job to work on the game full-time), expecting to build a whole new kind of game with their own four hands and no budget to speak of.

Before they could build the game, they had to build a programming language in which to write it. They spent more than two years constructing what they called ABL (for "A Behavior Language"), which encodes and controls virtual actors. "The actors' minds are written in ABL," Mateas explains. ABL itself has a sort of mind: enough artificial intelligence to decide how a particular character might, for example, simultaneously mix a drink, walk across the room, and yell at her husband, as a human actor could do.

That done, they built, again from scratch, another piece of AI, which they call a drama manager. It is a sort of artificial dramaturge and director, which looks at what the player and characters are doing and makes plot and dialogue choices intended to ratchet up and then release dramatic tension. Then they built a natural-language engine, which "listens" to what the player types in, looking for emotional and dramatic cues that the in-game characters can react to.

The game by now packed massive amounts of experimental technology under its hood, but what would it be about? They needed to create an intense drama in a confined space and with only a few characters. Influenced by Edward Albee's play *Who's Afraid of Virginia Woolf?* and also by several movies (Steven Soderbergh's *sex, lies, and videotape,* Woody Allen's *Husbands and Wives,* and Ingmar Bergman's *Scenes from a Marriage*), they decided to drop the player into a marital crisis. They hired actors to record five hours of dialogue, raw material from which the drama manager would build twenty minutes of game play.

In the end, they accomplished, they reckon, about 30 percent of what they had hoped to do. "We shot for the stars in hopes of getting to the moon," says Stern, "and we made it into orbit." In July 2005, standing together over Stern's computer in Portland, they pressed the button that "shipped," over the Internet, a new game called Façade.

When I set out to report this article, I thought I would bone up on video games and present myself as a suave expert. After all, I used to play a lot of Tetris. My aspirations to coolness lasted about three minutes, which was how long it took to load Electronic Arts' NBA Live 06. Jake Snyder, a twenty-something employee of the Entertainment Software Association, handed me the controls of a Microsoft Xbox 360 game console while two startlingly realistic basketball teams took shape before my eyes. As I stabbed at the unfamiliar buttons, I could barely control the ball. Flailing, I became aware that the game's color commentators were talking about . . . *me.* No, correction: they were *mocking* me. "Nice easy attempt, but they just can't make a shot," they said. "Totally disorganized," they sneered. I realized, face burning, that I had just lost the respect of a software product.

Determined to endure any further humiliations in private, I bought a copy of a critically acclaimed single-player game called The Elder Scrolls IV: Oblivion, a big hit from Bethesda Softworks and a new threshold of accomplishment in its genre. It came with a fifty-page manual full of instructions like this: "DISPEL: Removes Magicka-based spell effects from the target. Does not affect abilities, diseases, curses, or constant magic item effects. The magnitude of the Dispel must exceed the spell's resistance to dispel (based on its casting cost) in order to dispel it." I despaired. This sounded about as much fun as learning Microsoft Windows.

Entering the game, I was at first mystified and frustrated, but before long I was slaying goblins and pilfering valuables and casting spells and exploring caves. As the hours went by, I felt myself drawn in, then immersed, then reluctant to leave. I felt I was in the presence of a powerful medium, nothing like Tetris.

Oblivion's world is vast. A company spokesman told me I could explore for five hundred hours before seeing everything. The game enfolded me in lush, cinematic landscapes. It populated the cities, changed the weather, cycled through day and night. Looking down, I saw grass rendered in granular detail; looking up, I saw skies swept with feathery clouds; all around me I found innumerable creatures and towns and terrains. The illusion was magical.

But then it would all collapse. Approaching one of the characters, I would click for dialogue. The character would give a little canned speech introducing itself. In response to another click, it would mouth several bits of prerecorded dialogue. State-of-the-art games render action and environment with eerie realism and genuine aesthetic distinction. But their characters are dolls, not people.

It took me no more than a couple of minutes to see that Façade would be different. Grace and Trip, a married couple and old friends of mine, invite me over. He's blond, she's brunette, they seem to be in their thirties. As I arrive, I hear them arguing behind the door. After I knock, I'm cordially admitted by Trip into a small, sparsely furnished apartment with a view of towering apartment blocks glowing against a night sky.

Typing "Hi, Grace, you look great," I begin chatting with the couple. They try to draw me into their simmering argument, nudging me to take sides. I can say anything I like; there are no rules. I can be sullen and unresponsive (that got me kicked out of their apartment), or I can talk nonsense, but in most of my visits I try to behave like an improv actor, picking up on their lines and shooting back cues of my own — agreeing with one, criticizing the other, flirting with either or both. No two plays are identical. In a typical game, however, Grace and Trip will argue with each other, one may flatter me while the other questions my friendship, and the tension between them will build until feelings are raw and the story reaches a revelation or a breaking point. Here I'm playing as Ed:

TRIP: Okay, you know what, Ed, I need to ask you something.
GRACE: Trip —
ED: What?
TRIP: Grace, let me ask our guest a question. Ed, yes or no —
ED: Let him ask, Grace.
TRIP: Each person in a marriage is supposed to try really hard to be *in sync* with the other, right?
GRACE: *What?*

TRIP: I mean, when you're married, to make it good, you need to always be positive, and agreeable, and *together,* right?

ED: [Hesitates.]

TRIP: Yes or no.

ED: No, not always.

GRACE: What?! Oh, all right. Yes. Just admit it, Trip, admit it, we have a shitty marriage! We've never been really happy, from day one! Never, goddammit!

Here the drama manager is raising the tension to prepare for a revelation; notice how it demands my participation. The game can end in reconciliation or a split or, sometimes, neither. This time, Grace reveals that she let Trip stop her from becoming an artist, and Trip realizes his mistake, and they reconcile. "Ed, thanks for coming over," Trip tells me, his voice now subdued. "You — I think you helped us." I exit; the game is over. Next time something quite different will happen.

Façade won the grand jury prize at this year's Slamdance independent game festival and has drawn wide notice from industry journalists and bloggers. If you want to play it, you can download it for free, at www.interactivestory.net. So far more than 350,000 people have done so. Play and decide for yourself — but for me, playing Façade was both uncanny and frustrating.

Uncanny because Grace and Trip, despite being simply drawn, are at moments shockingly natural. "It was so subtle, was what impressed me," Will Wright, the prominent game designer, said when I asked him about Façade. "Most games beat you over the head with explosions and life-and-death situations and saving the world. And this is so subtle!" Trip, he marveled, can be slightly annoyed. "The fact that a character could be *slightly annoyed* in a game!"

Frustrating because for all their innovative AI-driven mechanics, Grace and Trip remain too dumb to sustain the illusion of humanness. When I played as a woman (I could choose my sex) and announced I was pregnant with Trip's child, Grace and Trip thought I was flirting with them. They really only guess at a player's meaning, and they don't guess very well. "It *kind of* works," says Doug Church, a respected designer with Electronic Arts, the 800-pound gorilla of U.S. video-game publishers. "It has moments of awesomeness. It has moments of *Wow, if I could play that, I'd be so excited!* But then you try the next step and bam! You hit a wall and the wrong thing happens."

Yet when it does work, when the game flows and the player has figured out how to collaborate with Grace and Trip, there *are* those moments. After a successful performance (to call it a game seems wrong), I jotted this note: "I feel a strange desire to please these characters and, despite my better judgment, touched when Grace reveals she's scared of painting and they reconcile." Façade feels like the small-scale, no-budget, first-try research project that it is. But it was still capable of working on my emotions.

In January, at Slamdance, Mateas and Stern met some investors who were excited about interactive drama. Many phone conversations later, they had a deal to raise $2 million for a commercial game. This was a crucial step for them. Stern, in particular, sees himself heading a commercial interactive drama studio. Both he and Mateas believe that today's video games occupy only a fraction of the potential market for interactive video entertainment.

"Most people — your sort of regular Joe or Jane on the street who loves television and movies — don't really get a whole lot out of games," Stern said when I asked who would buy interactive dramas.

"I think there's a real market for more character-rich, story-centered interactive experiences," Mateas added. "I think potentially it's a market that dwarfs the entire current video-game market. There is a huge untapped market for experiences that are not about action adventures, quests, killing monsters, and solving puzzles."

They have given their next game the working title "The Party." It is still in the conceptual stage, but they expect that, where Façade had two computer-generated characters, The Party will have ten, a far more complicated proposition, but dramatically richer. It will require not just two programmers but, once it enters production, ten or more. The graphics will be more detailed and polished. The action will take place in a larger space. The game will last about forty minutes rather than twenty. It will support more physical action, allowing the player to do things like rendezvous with characters in a private room, lock doors, carry things around, and fire a weapon. It will, they expect, understand the player better than Façade does, and support many more player moves.

And its aesthetic will be different. If Façade is a psychological drama, The Party will be a darkly comic social melodrama, along the lines of *Desperate Housewives*. In the prototype scripts, you find yourself cohosting a dinner party with your wife (or husband, if you play as a woman), who begs you to keep the conversation and liquor flowing smoothly. As guests arrive, the party fills with characters who have various designs on you and on each other. Your ex-girlfriend may try to break up your marriage; her angry husband may deck you; your neighbor may be snooping and your boss fishing for excuses to fire you. You can try to keep everyone happy, or you can hurl insults, or seduce your best friend's wife, or announce that you're gay, or refuse to admit guests (in which case your wife may let them in while shooting you angry looks), or lock your boss in the basement. You can try to mind your own business and be left alone. At every stage, however, the other characters — and behind them the drama manager — are conniving to draw you in. Madcap complications ensue.

There will be sex in the game, and there will be violence. There will be a gun, but only one bullet, so no shootouts. Here again the designers invert the conventions of Video Game Land, where shooting people is easy but talking to them is hard: in The Party, violence will be rare and dramatically meaningful, ricocheting through the game, as in life, with unforeseen consequences. Sex, likewise, will be dramatic rather than pornographic. It may disrupt a marriage or get someone killed. The sex will not be X-rated, but it will be realistic. "You may not literally see it, but the characters will be moaning," Stern said.

Mateas and Stern expect work on The Party to take two and a half years, at least. They hope to make the game a paying franchise and use the proceeds to push on toward their real goal: a game that understands natural language and generates its own drama.

The Party, like Façade, will assemble bits of prerecorded dialogue and preauthored plot points; the drama manager, as if stringing beads, will sequence the bits as it monitors the action. In the end, the game can be no bigger than its supply of prefabricated dramatic possibilities. The door to a world of truly open-ended drama will unlock only when a computer learns to write its own dialogue and plot twists, using rules that teach it to emulate a human playwright or screenwriter.

I raised an eyebrow. Can it be done? A simple prototype, Mateas said, is "totally doable within twenty years."

"We have every intention of doing those projects," Stern added.

*

The mainstream video-game industry is interested in hits, not research. On the business side of the industry, none of the executives I talked to had heard of Mateas and Stern, and the executives tended to regard the interactive drama project, when I described it, with polite skepticism, or — off the record — not-so-polite skepticism. "People love to blow shit up," one told me. He acknowledged exceptions, but said, "Blowing shit up is fundamental, because verbs are what make video games work. These guys are not going to succeed." At E³, I mentioned the Mateas-Stern project to Mitch Lasky, who himself has defied industry skepticism by making a fortune on cell-phone games. (He is now with Electronic Arts.) By way of response, he took a long drag on an imaginary marijuana joint. Good luck, was his attitude — but he wouldn't invest.

In the smaller world of game designers, by contrast, Mateas and Stern are a known commodity and are regarded with something like respectful curiosity. Designers have seen too many artificial intelligence failures to expect any kind of revolution, but at this point they would be happy if characters just got smarter. "A lot of people have worked on it," Doug Church, of Electronic Arts, told me. "Every year we're like, 'We're going to design incredibly intelligent, fluid humans who act realistically.' We try to take this huge step — and we fall all the way back down. At least," he said of Mateas and Stern, "they ended up somewhere new. It doesn't all work, but it is at least a step."

"It's a really hard problem, but it's one that we're incrementally going to solve," Will Wright mused when I asked him about creating believable characters. "It's a very tall mountain we're climbing." Mateas and Stern, he added, don't have the answer, but they have found a path uphill.

At the moment, all industry eyes are on a project of Wright's, one that enjoys EA's multimillion-dollar backing. (EA owns Wright's studio, Maxis.) Wright is nearing completion of a game called Spore, expected sometime next year. His last game, The Sims, was the biggest computer game hit of all time and a major innovation in its own right. Spore, as a feat of creative imagination and technical prowess, outdoes The Sims handily. It has enjoyed extravagant media hype for a game that has yet to ship a single unit. All I can say, having test-driven it, is that the hype understates the case.

Like Façade and The Party, Spore inverts traditional industry rules — but a different set of industry rules. Instead of outfitting the computer with a vast, prefabricated world for the player to explore, it leaves the designing of worlds to the players. But there is nothing, really, to "play": no need to win or compete. Instead, the player begins with a microbe, then helps it evolve into a creature of the player's own design. The creature spawns and becomes intelligent, eventually forming tribes and populating the planet; the player can then zoom out to explore a universe of planets and creatures, all created by other users and downloaded into his game from a mighty central server at Electronic Arts. In Spore, as Carl Sagan might have said, there are millions and millions of planets, all the fanciful, scary, inspired, or insipid handiwork of thousands or millions of players.

At E³, after watching Will Wright demonstrate the game to a couple dozen people in a small room with black walls, I was shown into an even smaller black room, where I sat down in front of an ordinary PC and went to work designing my own creature. To my astonishment, within five minutes I was comfortably building a scaly, beaked alien, as lavishly detailed and three-dimensional as anything one might see in a Pixar movie. Once I had given it enough body parts to move, it began . . . moving! It hopped. It walked. It made me giggle. Spore's most notable technical achievement is to teach the computer to animate whatever sort of creature anybody might design. Five legs? A buzz-saw-tipped tail and eyes astride the neck? No problem; the software, as if channeling Chuck Jones, looks at what you build and brings it to life, complete with characteristic movement, expressions, and even babies of the species. With not much more effort, I next terraformed a planet, giving it candy-colored mountains and icy lakes. It was as if I had a whole animation studio in my right hand.

Spore looks nothing like Façade and The Party. It is mainstream and big budget instead of independent and cheap, free-form in structure and timescale (you could play forever) instead of tightly woven and compact, visual instead of verbal (there are no people or words in Spore). It is, however, in some respects another bite from the same apple: born partly of frustration with the crippling limitations of existing video games, all three products seek to create a new audience for video game play by redefining the meaning of video game "play": *play* not as competition within rules (as in "play Tetris"), but *play* as creative fun (Spore is, at heart, a fantastically powerful toy) or *play* as dramatic performance (Façade and The Party are, at heart, interactive theater). Spore, if it succeeds, will evoke in the player a feeling of magical delight. Interactive drama, if it succeeds, will evoke emotional catharsis.

But how many consumers of entertainment actually want catharsis, especially after a long day at work? What most consumers of entertainment want is fun. The story goes that Will Wright was once approached by a designer who pitched a game that featured an elaborate new enemy system. As Heather Chaplin and Aaron Ruby relate the incident in their history of video games, *Smartbomb,* Wright heard out the pitch and then deflated the guy with one devastating sentence. "Hmm," he said, "that doesn't sound very fun."

Façade is ingenious, but it is not fun. It isn't really meant to be. The Party may turn out to be fun, even funny. But authoring fun is hard, and it is not obvious that interactive drama is a natural route to funness.

When the question of fun comes up, Mateas and Stern turn a little defensive. They are quick to say that games like Tony Hawk's Pro Skater, X-Men Legends, and Destroy All Humans! will always be with us, which is fine by them. They just want to do more. Mateas said, "When you go and see an intense movie or a seriously intense play, you don't walk out and go, 'God, that was fun!' It was a valuable experience and something you wanted to do and got something out of, but what you got out of it wasn't 'fun.' It was thoughtful, reflective, made you think about your own life, made you think about the human condition, moved you. And I think interactive media can do exactly the same thing, and potentially more powerfully than noninteractive media."

I asked what sort of aesthetic experience they had in mind. "Making players feel a true connection to characters on the screen," Stern replied. "You'd feel like you're immersed in an actual relationship with these characters."

"Yeah," added Mateas. "Having the player actually care about the characters."

They may be wrong about the commercial market for whatever they wind up creating, but they must be right about the human appetite for characters. A game, even a great game, is finished once played, but a great character, once met, lives forever. Think of

Sherlock Holmes and Mr. Spock, Don Quixote and Captain Ahab, Holden Caulfield and Humbert Humbert, Scrooge and Gandalf, Charlie Brown and Severus Snape.

In your mind, then, take the animation intelligence of Spore and the dramatic intelligence of Façade, increase their sophistication by orders of magnitude, and extend both vectors until they intersect. Imagine a game that could conjure a Holmes or a Spock, or that could create, or empower the player to create, all manner of original characters, each character not only animated but personified: *acted.* Imagine a game that not only conjured the cobblestones of Victorian London or the red sky of Vulcan but that charged each city, each planet, with a quantum of dramatic potential. Imagine, at last, entering those dramas and encountering those characters. Games, if such they were, might be as short as a sitcom episode or as long as a soap opera season; characters might be ones you created, bought, traded, or downloaded on a friend's recommendation; genres might span everything from comedy and fantasy to mystery and tragedy. You might not even need to choose: the software might watch how you play, learn your taste, and create dramas and characters and worlds to order. "Twenty years from now," Will Wright likes to say, "games will be as personal to you as your dreams, and as emotionally deep and meaningful to you as your dreams."

We can't know where the quest to build interactive drama might lead, but we do know that the dramatist's tools are the oldest and most potent of all emotional technologies. Sooner or later, drama will converge with the video game, the newest and most vibrant of all entertainment technologies. And then? Not long ago I attended a stage performance of Aeschylus' *The Persians,* the most ancient work in the dramatic literature. Even in translation and at a remove of 2,500 years, it left an audience of modern Americans feeling stunned and disembodied, as if the intervening millennia had disappeared. *Wow,* I heard myself think, *if I could play that, I'd be so excited!*

MICHAEL ROSENWALD

The Flu Hunter

FROM *Smithsonian*

ROBERT WEBSTER WAS in the backyard of his home in Memphis doing some landscaping. This was in the early winter of 1997, a Saturday. He was mixing compost, a chore he finds enchanting. He grew up on a farm in New Zealand, where his family raised ducks called Khaki Campbells. Nothing pleases him more than mucking around in the earth. He grows his own corn, then picks it himself. Some of his friends call him Farmer Webster, and although he is one of the world's most noted virologists, he finds the moniker distinguishing. He was going about his mixing when his wife, Marjorie, poked her head out the back door and said, "Rob, Nancy Cox is on the phone." Cox is the chief of the influenza division at the Centers for Disease Control and Prevention, in Atlanta. Webster went to the phone. He has a deep voice and a thick accent, which people sometimes confuse with pomposity. "Hello, Nancy," he said.

Cox sounded distressed. She told him there had been a frightening development in Hong Kong — more cases, and another death.

Oh my God, Webster recalls thinking. *This is happening. It's really happening this time.*

Some months before, a three-year-old boy in Hong Kong had developed a fever, a sore throat, and a cough. The flu, his parents thought. But the boy grew sicker. Respiratory arrest set in, and he died. The case alarmed doctors. They could not recall seeing such a nasty case of the flu, particularly in a child so young. They sent off samples of his lung fluid for testing, and the results showed that he did indeed have the flu, but it was a strain that had previously appeared only in birds. H5N1, it's called. Webster is the world's preeminent expert on avian influenza, and it was only a matter of time before the test results made their way to him. But he was not yet troubled. He thought there must have been some sort of contamination in the lab. H5N1 had never crossed over into humans. Had to be a mistake, he thought.

That was until Cox interrupted his gardening to tell him about the new cases.

It immediately occurred to Webster that he should be on an airplane. "I had to go into the markets," he told me recently. "I had to get into the markets as fast I could." He meant the poultry markets, where chickens are bought and sold by the hundreds of thousands. The little boy who died a few months before had been around some chickens, as have most little boys in that part of the world, where families often live side by side with their chickens, pigs, ducks, and dogs. If H5N1 was in fact in the markets, as Webster suspected, that was the beginning of his worst-case scenario: the virus could mutate in the chickens and perhaps other animals, and then acquire the know-how to pass from person to person, possibly initiating a pandemic that, he thought, might kill as many as 20 million people.

Webster has been predicting and preparing for such an event for his entire career as a scientist. His lab at St. Jude Children's Research Hospital in Memphis is the world's only laboratory that studies the human-animal interface of influenza. It was Webster who discovered that birds were likely responsible for past flu pandemics, including the one in Asia in 1957 that killed about 2 million people. He has spent a good part of his life collecting bird droppings and testing them for signs of influenza. Some of that collecting has taken place while he and his family were on vacation. One evening in Cape May, New Jersey, his school-age granddaughter ran toward him on the way to dinner saying that she had discovered some poop for him. He was so pleased.

A couple of days after Cox's phone call, Webster stepped off a plane in Hong Kong. He stopped at the University of Hong Kong to drum up some help to sample chicken droppings in the market. He also phoned his lab in Memphis and some scientists in Japan whom he had trained. He told them to pack their bags.

It occurred to Webster that there was a problem. The problem was H5N1. Neither he nor any members of his staff had ever been exposed to the virus strain, meaning they did not have any antibodies to it, meaning they had no defense against it. If they became infected, they would likely meet the same fate as the little boy who died.

They needed a vaccine. Four decades before, Webster had helped create the first widespread commercial flu vaccine. Until he came along, flu vaccines were given whole — the entire virus was inactivated and then injected. This caused numerous side effects, some of which were worse than the flu. Webster and his colleagues had the idea to break up the virus with detergents, so that only the immunity-producing particles need to be injected to spur an immune response. Most standard flu shots still work like this today.

Before they went to work in Hong Kong, Webster and his colleagues created a sort of crude vaccine from a sample containing the H5N1 virus. They declined to discuss the matter in detail, but they treated the sample to inactivate the virus. Webster arranged for a pathologist in Hong Kong to drip the vaccine into his nose and the noses of his staff. In theory, antibodies to the virus would soon form.

"Are you sure this is inactivated?" the pathologist said.

Webster pondered the question for a moment.

"Yes, it is. I hope."

And the fluid began dripping.

"It's very important to do things for yourself," Webster told me recently. "Scientists these days want other people to do things for them. But I think you have to be there, to be in the field, to see interactions." In many ways Webster's remarkable career can be traced to a walk along an Australian beach in the 1960s, when he was a microbiology research fellow at Australian National University.

He was strolling along with his research partner, Graeme Laver. Webster was in his thirties then, Laver a little older. Every 10 or 15 yards they came across a dead mutton-bird that apparently had been washed up on the beach. By that time the two men had been studying influenza for several years. They knew that in 1961 terns in South Africa had been killed by an influenza virus. Webster asked Laver, "What if the flu killed these birds?"

It was a tantalizing question. They decided to investigate further, arranging a trip to a deserted coral island off Queensland. Their boss was not entirely supportive of the adventure. "Laver is hallucinating," the boss told a colleague. They were undeterred. "Why there?" Laver once wrote of the trip. "Beautiful islands in an azure sea, hot sand, a baking sun, and warm coral lagoon. What better place to do flu research!" they snorkeled during the day. At night they swabbed the throats of hundreds of birds. Back at their lab, they had a eureka moment: eighteen birds had antibodies to a human flu virus that had circulated among people in 1957. Of course this meant only that the birds had been exposed to the virus, not that they were carrying or transmitting it.

To figure out if they were, Webster and Laver took subsequent trips to the Great Barrier Reef, Phillip Island, and Tryon Island. More swimming during the day, sherry parties at dusk, and then a few hours of swabbing birds. They took the material back to their lab at Australian National University, in Canberra. It is standard procedure to grow flu viruses in chicken eggs. So they injected the material from the swab into the chicken eggs, to see if the influenza virus would grow. Two days later the fluid was harvested. In most of the eggs, the virus had not grown. But in one of the eggs, it had grown. That could mean only one thing: the virus was in the birds.

Webster wanted to know more. Specifically, he wanted to know whether birds might have played a role in the influenza pandemic of 1957. He traveled to the World Influenza Center, in London, which has a large collection of influenza virus strains from birds and also antibody samples from flu victims. His experiment there was rather simple. He gathered antibody samples of several avian flu strains. Then he mixed the samples. What did the antibodies do? They attacked the bird flu strains, meaning the human flu virus had some of the same molecular features as avian flu viruses.

How could that be? The answer is something now known as reassortment. The influenza virus, whether it's carried by birds or by humans, has ten genes, which are arranged on eight separate gene segments. When two different influenza viruses infect the same cell, their genes may become reassorted — shuffled, mixed up. The net effect is that a new strain of flu virus forms, one that people have never been exposed to before. Webster refers to the mixing process as "virus sex." Perhaps Webster's greatest contribution to science is the idea that pandemics begin when avian and human flu viruses combine to form a new strain, one that people lack the ability to fight off.

After he entered the Hong Kong poultry markets, Webster needed only a few days to turn up enough chicken droppings to show that the H5N1 strain was indeed circulating. Along with many of his colleagues, he recommended that all the chickens in the market area be killed, to prevent spread of the virus. About 1.5 million chickens in Hong Kong met their maker. And that seemed to do the trick. The virus was gone.

But Webster had a hunch it would be back. The reason was ducks. Webster thinks the most dangerous animal in the world is the duck. His research has shown that ducks can transmit flu viruses quite easily to chickens. But while chickens that come down with bird flu die at rates approaching 100 percent, many ducks don't get sick at all. So they fly off to other parts of the world carrying the virus. "The duck is the Trojan horse," Webster says.

After the chickens in Hong Kong were killed, wild ducks probably relocated the virus to other parts of Asia, where it continued to infect chickens and shuffle its genetic makeup. When the strain emerged from hiding again, in Thailand and Vietnam in late 2003, it was even stronger. The virus passed directly from birds to people, killing dozens in what the World Health Organization has described as the worst outbreak of purely avian influenza ever to strike human beings.

Webster says the world is teetering on the edge of a knife blade. He thinks that H5N1 poses the most serious public health threat since the Spanish flu pandemic of 1918, which killed an estimated 40 million to 100 million people worldwide. Though the H5N1 strain has so far shown no signs that it will acquire the ability to transmit easily from person to person — all evidence is that flu victims in Vietnam and Thailand acquired the virus from direct contact with infected poultry — that has provided Webster no comfort. It's only a matter of time before this virus, as he puts it, "goes off." He has been saying this for several years. The world is finally taking notice. Elaborate plans are now being created in dozens of countries to deal with a pandemic. In November, President Bush requested that

$7.1 billion be set aside to prepare for one, with hundreds of million of dollars to be spent on further developing a new vaccine that was recently hatched in Webster's lab.

Webster has been advising federal health officials every step of the way. He does so out of fear of this virus and also because it is his job. When the H5N1 strain emerged in the late 1990s, the National Institute of Allergy and Infectious Diseases awarded Webster a major contract to establish a surveillance center in Hong Kong, to determine the molecular basis of transmission of the avian flu viruses and isolate strains that would be suitable to develop vaccines. "He's certainly one of those people in this field who have been way ahead of the curve in bringing attention to the issue," Anthony Fauci, the institute's director, told me. "He was out ahead of the pack. He's one of the handful of people who have not only been sounding the alarm but working to prevent this thing from turning into something that nobody wants to see happen."

Webster's job keeps him out of the country two to three weeks a month. Back in Memphis, his lab analyzes samples of influenza virus strains from around the world, to see how they are mutating. Recently health officials have reported finding H5N1 avian flu in birds in Turkey, Romania, Croatia, and Kuwait. It has not yet been found in birds in North America. If H5N1 makes its way here, Webster will likely be among the first to know.

This past June, I caught up with Webster at a meeting of the American Society for Microbiology, in Atlanta, where he was scheduled to deliver a speech about the threat of bird flu. There were more than five thousand microbiologists in attendance, which, because I am a recovering hypochondriac, I found strangely comforting. Walking around with Webster at a meeting of scientists is an experience that must be similar to walking around with Yo-Yo Ma at a meeting of cellists. When Webster walked by, people suddenly stopped speaking, a fact to which he seemed oblivious.

He opened his talk by asking a series of intriguing questions: "Will the H5N1 currently circulating in Vietnam learn to transmit, reproduce, from human to human? Why hasn't it done so already? It's had three years to learn how, and so what's it waiting for? Why can't it finish the job? We hope it doesn't."

He paused. "Is it the pig that's missing in the story?"

Webster explained that the strain is still not capable of acquiring the final ingredient needed to fuel a pandemic: the ability to transmit from person to person. For that to happen, Webster and others believe that a version of the human flu virus, which is easily transmittable between people, and the H5N1 avian virus have to infect the same mammalian cell at the same time and have virus sex. If H5N1 picks up those genes from the human flu virus that enable it to spread from person to person, Webster says that virtually nobody will have immunity to it. If an effective vaccine based specifically on that newly emerged virus isn't quickly available, and if antiviral drugs aren't also, many deaths will ensue.

Watching Webster speak, I couldn't help thinking that animals are not always our friends. It turns out that animals are a frequent source of what ails us. University of Edinburgh researchers recently compiled a rather frightening list of 1,415 microbes that cause diseases in humans. Sixty-one percent of those microbes are carried by animals and transmitted to humans. Cats and dogs are responsible for 43 percent of those microbes, according to the Edinburgh researchers; horses, cattle, sheep, goats, and pigs transmit 39 percent; rodents, 23 percent; birds, 10 percent. Primates originally transmitted AIDS to humans. Cows transmit bovine spongiform encephalopathy, or mad cow disease. In their 2004 book, *Beasts of the Earth: Animals, Humans, and Disease,* the physicians E. Fuller Torrey and Robert H. Yolken cite evidence suggesting that a parasite transmitted by cats, *Toxoplasma gondii,* causes schizophrenia. A couple of years ago, the monkeypox virus broke out among several people in the Midwest who had recently had close contact with pet prairie dogs.

And then there are pigs. For many years Webster has theorized that pigs are the mixing bowls for pandemic flu outbreaks. He has actually enshrined the theory in his house. He has a stained-glass window next to his front door that depicts what he perceives to be the natural evolution of flu pandemics. At the top of the glass, birds fly. Below them, a pig grazes. Man stands off to the left. Below all of them are circles that represent viruses and seem to be in motion. They are set in a backdrop of fever red.

The pig is in the picture because its genome, perhaps surprisingly, shares certain key features with the human genome. Pigs readily catch human flu strains. Pigs are also susceptible to picking up avian flu strains, mostly because they often live so close to poultry. If a human flu strain and an avian flu strain infect a pig cell at the same time, and the two different viruses exchange genetic material inside a pig cell, it's possible that the virulent avian strain will pick up human flu virus genes that control transmission between people. If that happens with H5N1, that will almost certainly mean that the virus will be able to pass easily from person to person. A pandemic may not be far behind.

During his talk in Atlanta, Webster pointed out that this H5N1 virus was so crafty that it has already learned to infect tigers and other cats, something no avian flu has ever done. "The pig may or may not be necessary" for a pandemic to go off, Webster said. "Anyway, this virus has a chance at being successful." He said he hoped world health officials "would keep making their plans because they may face it this winter. We hope not."

I went hunting with Webster. Hunting for corn. His cornfield is on a patch of land he owns about 5 miles from his home on the outskirts of Memphis. He grows genetically modified corn that he gets from Illinois. An extra gene component known for increasing sweetness has been inserted into the corn's DNA, producing some of the sweetest corn in the United States. Three of his grandchildren were with us, visiting from North Carolina. They had come, among other reasons, for Webster's annual Corn Fest, where members of the virology department at St. Jude Hospital gather in his backyard to sit around eating corn on the cob. The record for the most ears of corn eaten in one sitting at the Corn Fest is seventeen. The record-holder is the teenage son of one of Webster's protégés. Webster reports the prize was a three-day stomachache. He encouraged me not to beat this record.

"There's a good one," Webster said, bending down to pull off an ear. He was wearing long shorts, a plaid blue shirt, and a wide-brimmed canvas hat. He had been fussing around among the stalks for a few minutes before he found an ear he liked. He seemed unhappy with the quality of the corn, muttering into his chest. In between picking some ears, I asked why he was down on the crop. "I believe I planted too soon," he said. "The ground was still too damp." This caused many of the ears to bloom improperly. I asked why he had planted so early. He said, "I had to be in Asia." It occurred to me that attempting to stop a global epidemic was a reasonable excuse for a so-so batch of corn.

Webster was home this weekend for the first time in many weeks. He had been in Asia and back nearly a dozen times in the past year. I asked Marjorie Webster how often she sees him, and she replied, "Not much these days." It is a sacrifice she seems willing to make; Webster has told her plenty about the bug and what it can do.

We picked corn for about half an hour, then went back to Webster's home to do some shucking. He shucked at a pace nearly double mine. We must have shucked 250 ears of corn. We placed the shucked ears in a cooler of ice. By noon we had finished, so I decided to go do some sightseeing. Beale Street, Elvis impersonators, several barbecue joints. A little before 5 P.M., I wandered into the lobby of the Peabody Hotel, a landmark. I wanted to see the ducks. Since the 1930s, ducks have swum in a fountain in the hotel's lobby. The ducks live upstairs in a sort of duck mansion. In the morning they ride down in an elevator. When the elevator doors open in the lobby, the ducks wobble down a red carpet, single file, about 30 yards, in front of hundreds of people who snap photographs as if they were duck paparazzi. When the ducks plop into the fountain, people cheer. At 5 P.M., the ducks are done for the day; they wobble back along the carpet to the elevator, then ride back to their mansion for dinner. One generally has to witness the occasion to believe it.

I wondered whether Webster had ever tested these ducks. That evening, at the corn party, after my third ear and Webster's second, I told him that I had gone to see the ducks. "Oh, the Peabody ducks," he said, the first time I'd seen him visibly happy in days. "The kids loved the ducks when they were little." I asked whether he liked the ducks too. "Why not? I enjoy the ducks," he said. I said, "Have you ever swabbed them?" He answered, "No. Sometimes you just don't want to know. There are some ducks I won't swab."

BONNIE J. ROUGH

Notes on the Space We Take

FROM *Ninth Letter*

THE WOMB is the smallest space in which a human being may live.

Most human babies, when they are born, weigh from 7 to 8 pounds — about the same as a gallon of milk.

When I was born, I weighed 7 pounds, 8 ounces. When my mother was born, she weighed 4 pounds. She lost 2 pounds before she began growing again. When my mother's mother was born, it didn't cross anyone's mind to weigh her.

Little babies — under 5½ pounds — are said to be much more difficult to care for than big babies.

When I stand as tall as I can, I am 5 feet, 8¾inches high. I most often weigh 142 pounds. Buster Keaton weighed that much, and so did a world-record-setting paddlefish caught in Montana in 1973. My weight is also the weight of a newborn giraffe, which stands 6½ feet tall the first time it takes to its feet, casting a much bigger shadow than I do.

Hermit crabs grow out of their shells — not into them.

Our little blue tent has 37 square feet of floor space — a little over 5 feet by 7 feet. It is a two-man tent. But we are one man and one woman. Still, it fits us snugly, with enough arch to stretch up an arm, and a taut wall for our toes to brush in sleep.

Hermit crabs drag their shells around a spring of fresh water on the bank of grass behind the palm trees, near our little blue tent. They hiss through the low tropical forest, among fallen fruits. They hiss onto the beach, shells scraping, making tracks crossing tracks crossing tracks.

The hermit crab, which is not a true crab, has a somewhat stiff exoskeleton but lacks a protective carapace. This leaves exposed his tender belly. To protect himself, the crab looks for snail shells, abandoned or soon to be abandoned. When he finds one he can back snugly into, he turns his small right-hand claw backward to grip the spiral inside his shell, hoisting the whole thing onto his back. He puts it down when he is resting or eating, and heaves it up again when he takes a walk. The hermit's left-hand claw is enormous. He uses it to gather food and fend off attackers. The hermit crab also uses his large claw as a front door, slamming it across the entrance to his shell when he wants to be alone.

Hermit crabs prefer a high volume-to-weight ratio when selecting a shell to live in. A light, roomy shell is ideal. A hermit crab will look for a new shell when his old one becomes dirty, broken, or otherwise unsuitable — or he will change shells for no apparent reason at all. When a hermit crab molts — sometimes more than once a year — he sheds his old skin. It peels off and slides over his head like a full-body disguise. The crab then experiences a few days of terrifying nakedness, when his new skin is soft and supple. Finally the new skin hardens. Only during his few days of soft-skinned vulnerability may a hermit crab grow. When he emerges from his molting hole in the sand — if other crabs haven't already found him and eaten him alive — he searches out a bigger shell for his newly bigger self.

Scientists and activists note a dire situation on the world's beaches: a shortage of snail shells for hermit crabs to move into. One way for a hard-up crab to find a new shell is to select one from someone else's back. "Shellfighting" ensues — one crab twisting the other out of his shell, and the evicted crab scrambling for cover. Scientists have observed that when a shiny, roomy new shell is placed on the beach, a crowd of hermit crabs will converge upon it. But instead of mayhem, a ritual of great civility follows. The crabs arrange themselves in order of size. The largest crab exits his shell and climbs into the shiny new one. The crab just below him in size takes his old shell, and so on down the line, until every crab, down to the teensiest, has a comfortable new home. It is clear to everyone involved that the teensiest crab is pleased with his take, and would be miserable in the largest shell — perhaps even unable to lift it. In his case, the ideal home sits right up against the belly with a smooth plate of calcium carbonate, and has a doorway no larger than his own left hand.

Under duress of a snail shell shortage, hermit crabs have in recent years been sighted carrying plastic pill bottles and airplane liquor bottles on their backs. Elizabeth Demaray, a New Jersey artist, proposes a rescue operation: the distribution across American beaches of custom-molded plastic homes for hermit crabs. Since pollution and consumer excess are blamed for the decline of shoreline health and the resultant tumble in snail populations, the artist suggests that corporations be made to pay for the plastic hermit-crab homes (which have an irresistibly high volume-to-weight ratio). In honor of their support, these corporations would have their logos printed on the plastic.

More than 1 million Americans call a recreational vehicle their primary residence.

There are no reliable statistics for the number of Americans who live in cars and trucks. But people who live in cars and trucks by choice say they like the convenience of a cozy, warm place to read and then sleep, even if it's hard to host a party. People who live in cars and trucks by necessity are less often interviewed.

Cars, trailers, tents, yurts, gers, teepees, wickiups, wigwams, and benders are houses you can pack up and take with you. But sometimes houses move from place to place only in the mind. For some Inuit people, the work of building an igloo was a daily task, like building a fire. A new evening, a new landscape, a new house. These were houses not meant to be lived in more than once and certainly not by more than one family.

Some people find ways to bed down away from the land entirely. Some live in boats on water, some sleep in airplane cabins, and some astronauts subsist in weightless space.

For kids, NASA uses an apple metaphor.

"Pretend your apple is the planet Earth. It is round and beautiful. It is full of good things. See how its skin hugs and protects it? Cut your apple into four equal parts. Three of those parts are water on Earth. You may eat these three parts. The fourth that is left is dry land. Cut that dry land part in half. One part is land that is too hot or too cold. Eat this part. Cut a little more than one half. Eat the smaller part. It is too rocky or too rainy. Food can't grow on this part. Peel the skin from this part. Eat the inside. Look at the little part that is left. This shows us all that we have to grow food for the world. Some of this little bit has houses, schools and malls on [it]."

Earth provides about 58 million square miles of land. Of that land, 23 million square miles is considered habitable by human beings — as long as food can be brought from elsewhere.

The average American home has 2,200 square feet of livable space — a little more than a singles tennis court. The average American family consists of 3.86 persons. The average home space an American inhabits is 570 square feet — the size of the two rectangles into which tennis players must serve.

Before we married, my husband and I lived in a house of 550 square feet. When we married, we moved into a house of 1,175 square feet. When we finished graduate school, we moved into the city, into a house of 1,800 square feet.

Unlike hermit crabs, my husband and I have given ourselves room to grow in our new house. If we don't fill it with the work of our brains and hands, we will create children to share our space with us, thereby decreasing the relatively high space-to-human ratio in our home.

Biltmore Estate, in Asheville, North Carolina, is America's largest home. It has 250 rooms, including 34 bedrooms and 43 bathrooms. It has almost as many fireplaces as bedrooms and bathrooms combined. Tycoon George Washington Vanderbilt finished building the Biltmore in 1865, when it had 4 acres of floor space — enough for three football fields.

In terms of Earth's geologic history, space was not a gift to life but vice versa.

Fed up with oversimplified guidelines for what size tank will fit aquarium fish, scientists G. J. Reclos, A. Iliopoulos, and M. K. Oliver offered new formulas in the June 2003 issue of *Freshwater and Marine Aquarium.* "The best known [rule] is the 'one liter of water per cm of body length' (or 'one gallon of water per inch of fish') rule which, in our opinion, is useless . . . The statement that shows the impracticability of this rule is 'Twenty neon tetras each measuring 1.5 cm may fit a 20 liter tank but one 30 cm fish will not.'"

In place of this shoddy guideline, the authors provide the following (along with easy formulas for tank height and width), where L=tank length and FL=maximum expected fish length:

Peaceful fish, peaceful tankmates:
L = FL × 4
Mild temperament, cruise predators:
L = FL × 5 × 1.3
Aggressive fish, good swimmers:
L = FL × 5 × 1.2
Aggressive fish, poor swimmers, ambush predators:
L = FL × 5
Aggressive fish, cruise predators:
L = FL × 5 × 1.5
Pair of aggressive fish, cruise predators:
L = FL × 5 × 1.5 × 1.2

There is a range of literature on the proper way to design zoo enclosures. Some guidelines say that most mammals need an enclosure that allows them to explore changing landscapes, work for food, search for mates, get away from the viewing public, and simply investigate their territory. On the other hand, at least one scientist states that in designing zoo enclosures, keepers need only bear in mind that "animals will usually only move to find food, escape danger, or find mates."

I have seen six smooth belugas negotiating space in a single aquarium pool, raving polar bears pacing zoo pens, and caged birds wedging their beaks between bars.

Belugas are the only all-white whales, colored to blend with the edges of Arctic ice fields, where they dwell all year. Gray whales migrate much farther — up to 12,500 miles every year — but at home near the ice, belugas dive a thousand feet deep.

Polar bears move in enormous, private northern arcs. They wander thousands of miles every year on paths programmed inside their brains. At least one bear has been tracked pacing the ice 3,000 miles from Alaska to Greenland — then back.

The Arctic tern migrates between the North and South Poles, a journey of 10,000 miles twice a year, perhaps experiencing more sunlight in its lifetime than any other species on Earth.

Clipping a bird's wings, obviously, does nothing to diminish its impulse to fly.

The urge to migrate is deeply rooted in human ancestry.

I was born in Washington State and grew up 70 miles from my maternity ward. When I was three, my parents and I moved from a one-story house to a two-story house a mile away. When I graduated from high school, I moved across the state. When I graduated from college, I moved across the country. By this time I had discovered flight. I traveled to a new continent every year, and wore out many pairs of shoes.

For over 99 percent of human history, we moved as nomads: with the seasons, in small groups, to the places most likely to provide edible plants and game.

Travel writer Bruce Chatwin loved the idea of human migration. In 1970, he wrote the following in "It's a Nomad, *Nomad* World":

> Some American brain specialists took encephalograph readings of travellers. They found that changes of scenery and awareness of the passage of seasons through the year stimulated the rhythms of the brain, contributing to a sense of well-being and an active purpose in life. Monotonous surroundings . . . produced fatigue, nervous disorders, apathy, self-disgust and violent reactions. Hardly surprising, then, that a generation cushioned from the cold by central heating, from the heat by air-conditioning, carted in aseptic transports from one identical house or hotel to another, should feel the need for journeys of the mind or body, for pep pills or tranquillisers, or for the cathartic journeys of sex, music, and dance.

Chatwin also said it simply. "The best thing is to walk."

My neighbor is a retired mail carrier. He takes a three-hour walk every morning, returning in time for lunch. Then he takes a three-hour walk every afternoon. He walks in all weather, under all manner of clothing. Puzzled by his daily miles, I asked if he misses delivering mail. No, he said. He misses movement.

People who seem small end up in small spaces.

The best way to shelter a child is against the body. But humans who do not enjoy wearing infants against their bodies, and who do not want their infants to get away, often store their babies in small enclosures, such as baskets, plastic carriers, cradles, bassinets, playpens, and cribs.

Some people set up house in cardboard boxes. Sometimes they are children in trim backyards, and sometimes they are grownups on public property.

My grandmother has just moved from her two-story home overlooking Puget Sound into a room at Merrill Gardens at Mill Creek. She had to pass a test to qualify for lower monthly rent. She had to prove she wouldn't require too much assistance.

I keep telling my grandmother that her new home is going to be like college, or nursing school, all over again. Friendly faces in dorm rooms up and down the hallways. She seems only partly convinced, perhaps because dropping out means something different now.

Tiny people, huge people, hairy people, albino people, legless people, and two-headed people could all be viewed for the price of a coin in the U.S. between 1840 and 1940, when freak shows were wildly popular. Lots of times the freaks were displayed in cages or other small enclosures, but it seems they were released after the show.

Judges in many U.S. states have declared that a prison cell should allow no less than 60 square feet of space per prisoner, but this edict is rarely followed in U.S. prisons, which are endemically overcrowded. The American prison at Guantánamo Bay features prison cells of approximately 48 square feet — about 6 inches larger on all sides than our blue tent. Humans have been — or are — stored in much closer quarters: hulls of ships, forbidding camps, airless train cars lurching in the night.

It would take 3.4 billion Biltmore Estates to swallow Earth's livable space. It would take 294 billion average American homes, or 13 trillion Guantánamo cells. Earth's population ticks upward every second. At the time of this writing, the planet supported about 6.55 billion human lives.

Perhaps, then, it should be no surprise that humans commonly discuss the colonization of Mars, the moon, and other planets in the solar system.

Claustrophobia is the human fear of small or enclosed spaces.

Rose Hill Cemetery in Whittier, California, is the largest cemetery in the United States. It is the size of 106 football fields. Only a handful of people with claims to fame are buried here — a few professional athletes, a few musicians — and there is space available for newcomers.

Less then half the size of Rose Hill Cemetery, Arlington National Cemetery is the second largest burial place in the United States. Former presidents, astronauts, and Supreme Court justices are buried there. Arlington has some space available, but to be buried there, one must have been a U.S. president, a prisoner of war or other unusual veteran, or their spouse or child.

When there is no more room for the dead, or when we try to hide the dead, humans bury bodies one atop another. When there is no more room for cemeteries, humans build homes and markets and parking lots upon them. In Alexandria, Egypt, for example, archaeologists were recently able to dig down into two-thousand-year-old Necropolis, a city of hundreds of tombs, only when a building was torn down to make room for a highway. After a whirlwind excavation, workers sealed the tombs over again.

Despite the inevitable tight fit, thousands of people every year purchase small plots of land for their bodies, or little lockers for their cremated remains.

Some traditions allow corpses to feel the freedom of space. They place their dead in trees, or slip them into rivers, lakes, and oceans.

My grandmother has asked us to check the "discard" box on the funeral parlor's form when she dies. She will not have a coffin or an urn.

In the summer of 1926, escape artist Harry Houdini bought a bronze coffin and first performed a new trick. Stepping inside, Houdini bade his assistants close the lid and lower him into a pool of water. Ninety minutes later, the assistants pulled up the coffin. The magician emerged, smiling. But Houdini was able to perform this trick only a few times. A stomach illness killed him that fall, and he was buried in the bronze coffin.

A standard coffin is 80 inches long and 23 inches wide. The largest regularly produced coffin available to U.S. funeral homes is 88 inches long and 36 inches wide. It is capable of respectfully accommodating a 500- to 600-pound corpse.

The tiniest coffin available is made for premature infant deaths. It is 10 inches long and 5 inches wide.

ROBERT M. SAPOLSKY

The Olfactory Lives of Primates

FROM *The Virginia Quarterly Review*

Dear Chris,

I'm sorry you couldn't make it to Brad and Caitland's wedding. It was pretty good. The beginning dragged — the usual, everyone standing around, sniffing each other's breath, figuring out who was from Caitland's family, who from Brad's. You could see people getting confused by Jessica, Caitland's niece, the one who's adopted.

At one point, this very attractive second cousin of Brad's came up and sniffed my butt. Okay, I'll admit it, it made my day. Then these two drunk guys from Caitland's office got into this pissing contest, completely soaked a tablecloth doing it. I don't know why they bothered — one of the guys had no testes, and he was terrified of the other guy — majorly acrid sweat.

Totally poignant with Hugh, Caitland's dad. She didn't think he'd last for the wedding. You could smell the cancer from across the room. I've always liked the guy.

One really funny thing — Caitland's sisters were both ovulating (it occurred to me that Brad had done that to them, which would have been really crappy of him), and they kept sniffing Tom's crotch. They were trying to make it seem like this big joke, but it was obvious that they were, like, totally serious. Lame.

SCIENTISTS ARE CAPABLE of being as childish as anyone else, and fights often erupt over what name to give to some new discovery. This happened in the middle of the last century and concerned the name for a part of the brain. Deep in the brain's underbelly, far from that gleaming cortex doing string theory physics, is this interconnected cluster of ancient regions, like the amygdala, hippocampus, and hypothalamus. The old guard of neuroscience had long called this network the rhinencephalon — "nose brain."

This made sense. In a rat, there's this gigantic mass of brain cells that detect odors, and it funnels all that olfactory information, by way of cable-like "projections," into that amygdala/hippocampus/hypothalamus network. Meanwhile, a bunch of young Turk neuroscientists ignored this rhinencephalon label. They were analyzing what the amygdala, hippocampus, etc., actually did. For example, what behavioral changes would occur if one of those regions was damaged? And they were finding some very interesting effects — changes in sexual or aggressive or maternal behavior. These folks concluded that this network was about emotion (and for reasons I've never figured out, they called this network the "limbic" part of the brain).

Rhinencephalon or limbic system? Smell or emotion? Coke or Pepsi? After factional violence that left thousands dead, a kumbaya-esque solution became obvious. For your typical mammal, there was no conflict — emotions equal odors. Sexual, aggressive, or maternal behavior never occurs outside the context of olfaction.

Olfaction is a unique sensory modality for a mammal. To appreciate this, we have to consider how neurons — the main type of brain cell — communicate with each other, often in long sequences of cells. Neuroscientists play a game of counting how many neurons it takes to go from Brain Region A to Region B. If B is only one neuronal connection away from A, A is likely to have a big influence on B's function. But if A has to send its message snaking through a line of ten neurons to reach B, A's not going to be of much consequence to B.

For sensory systems, how many steps does it take to get from the eye, ear, or patch of skin to that emotional limbic system? Roughly ten steps. Take vision. First there's a layer of neurons in the cortex that breaks the visual scene into dots, then a next layer turning the dots into lines, then collections of lines, on and on. Finally, an Ice Age later, by a neuron's temporal standard, visual information trickles to the limbic system, and you activate an emotional response appropriate to seeing the face of someone intent on, say, seducing you or ethnically cleansing you. All of the sensory systems, that is, except olfaction. How many steps from smelling something to the limbic system? Just one.

The result is an intertwining of the limbic and the rhinencephalonocentric worldviews. Most mammalian species give off odors —

pheromones — containing all sorts of information relevant to emotion in the recipient. An animal's pheromones communicate its genetic pedigree, gender, approximate age, if it is ovulating, if it has testes, if it is sick, happy, or terrified. Just consider — a dog sure doesn't want an opponent to know that he's scared. But he can't will his anal scent glands not to exude the stress-hormone-laden pheromones that broadcast his terror. But he can sure try to put a lid on those scent glands — by tucking his tail between his legs.

All of that is the type of information that can change emotions in the recipient, and in most mammals, more readily than information from other sensory systems can. Olfaction can also shape bodily responses that accompany particular emotions. Get the right pheromonal information looping to the hypothalamus, and you nudge it toward increasing your heart rate. There are certain pheromones released by rodents that will change the likelihood of ovulation, others that will change when puberty occurs. Even certain pheromones that make a pregnant female miscarry.

And olfaction has easier access to memory than the other senses. One of the key structures related to memory formation and retrieval is the hippocampus (ponder the significance of the fact that a key portal to memory sits in a brain network most relevant to emotion). And, unlike those meandering projections from the other sensory systems, olfaction fast-tracks its information to the hippocampus, with fewer intervening steps. As a result, a rat can learn an olfactory task faster than we can learn some visual ones — trashing that business about us sitting at the top of a pyramid of species complexity.

So olfaction rules. Certainly in a rat, which devotes an outlandish 48 percent of its brain to olfactory processing. Rhinencephalon indeed. Same olfactory obsession in dogs, where you take the dog out for a quick walk in a drenching thunderstorm, and he has twenty spots in the neighborhood to sniff in order to catch up on gossip. There are even dogs who are professional noses — those bloodhounds trained to smell out explosives and drugs and anti-Dubya writing in suitcases. Olfaction is even a big deal for nonhuman primates. Much less of the brain is devoted to olfaction than in a rodent, with an emphasis instead on vision — for example, the evolution of sophisticated stereoscopic color vision for spotting a smidgen of a color signaling the right kind of ripe fruit, amid a riot of rain-forest foliage. But that emphasis on vision only counts for so much. Troop of baboons, someone gave birth during the night, new mom is now on the ground, holding the baboonito, and everyone comes to check it out, to ask the same question that we all ask — boy or girl? And they don't spread the kid's legs to look. They spread them to sniff.

So what about humans? We're more olfactory than most would guess. After all, we spend fortunes on perfumes and Dainty Floral Country Garden sprays to use in the bathroom after we've really cut one. We do the olfaction/hormone business familiar to other mammals. For example, there is that tendency of women roommates to synchronize their menstrual cycles, and this is accomplished through pheromones. Or that women can smell (but not be consciously aware) if some guy has his testes. Or that men can smell (again not consciously) whether someone is ovulating. (High-tech science: get some women volunteers at known points in their cycles to rub their undeodoranted armpits with a cotton swab, stick it in a Mason jar, and let guys sniff the insides of various jars. And not only do guys tend to distinguish the smell of the ovulatory cotton swab from the others, they think it smells *better.*) And that olfaction/memory linking works in humans: this accounts for that Proustian moment where an unexpected odor wafts by and suddenly you're transported back to kindergarten with this wild intensity, back amid the primary-colored plastic chairs and the Elmer's glue. Olfaction does things to our human heads that the other sensory modalities can only dream of.

But despite those amazing human olfactory abilities, we still have a very atrophied olfactory system, compared to other mammals. We can't remotely distinguish odors as well as most other species. We devote only 1 percent of our brain mass to olfaction. But here's the measure that really amazes me.

A truly unsettling finding came a few decades back that chimps and humans share 98 percent of their DNA. Ninety-eight percent. This is boggling. So this naturally leads to the question, what's the 2 percent difference about? The answer came recently. A few years back, that mammoth public works venture, the Human Genome Project, decoded the human genome, producing a billions-long sequence of letters that comprises all our DNA. And then, a few months back, the chimpanzee genome was revealed. And at last we get to see what that 2 percent difference is.

And it turns out to be weird. Not a thing having to do with our brains working so differently than theirs — no genetic explanation for literature, art, megastates, termite sticks, the bunny hop (which increasingly strikes me as logical). Our "me-ness" is not all that anchored in our genes. There were some genetic differences concerning body hair. Others about immune functions — chimps handle malaria better than we do, we handle tuberculosis better than they do. Some about reproduction, making it unlikely that there's some human-chimp hybrid out there. But the biggest difference concerned olfaction. When an odor — a molecule that has floated off, become airborne, from sweat, from a mound of cinnamon or a rotten egg or pollinating flower or exhaust pipe — reaches your nose, it binds to a subset of literally thousands of different olfactory receptors which, when activated in particular patterns, send a *Guess what I just smelled* message to your brain. Those receptors are coded for by genes, and it turns out that half — *half* — of the genetic differences between chimps and humans concern olfactory receptors. They've got them, and we've functionally disabled ours into what are called pseudogenes. As we split off from our last common ancestor a few million years back — an ape already with an atrophied olfactory system — the most common genetic shift that would ultimately differentiate us from chimps was that we decayed into having a lousy sense of smell.

And what are the consequences? We have become disproportionately specialized in our other senses. We can be aroused by an erotic picture. We can be moved to tears by music. We can read this essay in Braille.

But even our nonolfactory senses are not so hot. Raptors see better than we do. Nocturnal animals hear better. Our senses aren't great. But our thinking about our senses is amazing. You can decide whether some curry tastes the same as the one you made last July. You can remember the sensation of someone's hand moving down your body. Or you can sit alone in a loft with a stack of empty sheets of music paper and imagine what the symphony will sound like. We have been freed from the concrete here-and-now of sensation — as we have been from the here-and-now of emotion, of thought, of everything. And while that may make for less interesting wedding parties, it sure is central to what makes us human.

JOHN SEABROOK

Ruffled Feathers

FROM *The New Yorker*

ON A WINTRY EVENING in January, the Smithsonian threw a book party in the Castle — a turreted folly on the Mall, in Washington, D.C. — to celebrate Pamela Rasmussen's monumental new work, *Birds of South Asia: The Ripley Guide,* which had recently been published, in two volumes, by the Smithsonian and Lynx Edicions. The book, illustrated by John Anderton and other artists, puts the highest standards of professional ornithology into a form that an amateur can use in bird watching, bridging a schism between professional and amateur bird-lovers that has existed for almost a century. Rasmussen examined everything that is known about birds in India, Bhutan, Bangladesh, Nepal, Sri Lanka, Pakistan, Afghanistan, and the Maldives — she measured, described, and plotted range maps for most of 1,441 species, serving as sole judge in hundreds of difficult decisions about which birds to include on the final list. Along the way, she negotiated many other obstacles, including the death, in 2001, of S. Dillon Ripley, a grand old man of American ornithology, who was the book's originator and guiding spirit. Most spectacularly, her research helped lead to the unraveling of the greatest ornithological fraud ever committed — a convoluted skein of theft and data falsification that was perpetrated by the late British ornithologist Colonel Richard Meinertzhagen.

At the party, Rasmussen mingled shyly with the clubby, moneyed world that used to dominate natural history and still lingers in ornithology. David Challinor, Harvard '43, now eighty-five years old and a senior scientist emeritus at the Smithsonian, greeted her by saying, "Marvelous, Pam, marvelous. You stayed with it." Rasmussen smiled and nodded, a bit stiffly. She is keenly aware of the differences between her own upbringing, in Oregon, and that of the East Coast scientist-aristocrats she has come to know at the Smithsonian. Rasmussen is forty-seven years old and stands six feet tall in her high heels. She has Nordic features and wide-set pale blue eyes. She sometimes appears startled when you address her, as though she'd been thinking about something else. As she made her way around the room, colleagues from the Smithsonian praised her extraordinary ability to notice tiny details about bird specimens that other researchers miss, a condition Rasmussen herself diagnoses as "attention surplus disorder." Bruce Beehler, a former coworker, said, "Very few younger researchers have Pam's ability to stay focused on one thing, one specimen, for hours and hours. They need ten things happening at once."

Conversation returned several times during the evening to Colonel Meinertzhagen, who died in 1967. Meinertzhagen enjoyed a formidable reputation in international ornithology. He was a chairman of the British Ornithologists' Club and the recipient of a Godman-Salvin Medal, one of the highest honors in British birding. His unorthodox methods and surprising finds had been the subject of rumor during his lifetime, but there was never any substantiated accusation of fraud. Three largely adulatory biographies of Meinertzhagen have been published since his death — one by a soldier, one by a professional game hunter, and one by a birder. *Duty, Honor, Empire* (1970), by John Lord, and *Warrior* (1998), by Peter Hathaway Capstick, don't say anything about fraud. Mark Cocker's *Richard Meinertzhagen: Soldier, Scientist and Spy* (1989) does consider many of the rumors, but ultimately rejects them. It wasn't until more than thirty years after his death, when circumstances brought Meinertzhagen to the notice of an unusually attentive researcher, that the extent of his deception was revealed.

One day in 1967, when Rasmussen was eight, her father, Dr. Chester Murray Rasmussen, came into her room, glanced contemptuously at the dolls she and her younger sister, Sally, were playing with, and said, "Why don't you girls get interested in something useful?" Her father was an osteopath, and Pam remembers him as "the biggest man you had ever seen." He spent a lot of time in his basement den, which he had decorated with the heads of animals that he had killed on hunting trips. (A few years later, Dr. Rasmussen deserted the family.) Pam's mother, Helen, a strict Seventh-Day Adventist, let the children know that she disapproved of what he was doing down there. "Sinning," was how Pam understood it.

A couple of days after that encounter, Pam's mother bought her the junior edition of Oliver Austin's *Birds of the World,* illustrated by Arthur Singer. Pam had never thought much about birds before, but she quickly became obsessed with the pictures in the book. "I just thought the way the birds looked was so wonderful," she said. She tried to get her sister interested too. "Pam would open the book and say, 'Okay, which of the birds on this page do you like best?'" Sally told me. "Then she'd do the same thing for the next page — all the way through the book."

After poring over the illustrations, Pam went out to the marsh behind the house, which was situated about 20 miles west of Portland, and recognized a bird — a long-billed marsh wren. "Now that I knew its name, it was thrilling," she said. She had bird-watching birthdays at the Oregon beach, in October. "There we'd be, with Pam, freezing, looking for ducks," Sally, who is now a financial writer, said. For gifts, Rasmussen always wanted bird books. (If it was an expensive book, like the first edition of *Parrots of the World,* by Joseph M. Forshaw, it had to cover two or three gift opportunities.) In addition to looking at the pictures, Rasmussen studied the text of these books carefully — except for the chapters entitled "Fossil Birds." Her mother told her not to read those.

At Walla Walla College, an Adventist institution in Washington State where Rasmussen spent six years, earning her master's in biology in 1983, evolution was not taken seriously. "Except," she said, "when we learned what was wrong with it." It wasn't until Rasmussen reached the University of Kansas, where she did her dissertation on cormorants in Patagonia, receiving a Ph.D. in 1990, that she studied Darwin in depth. "It wasn't like the scales fell from my eyes," she said. "You don't really need Darwin to be interested in bird diversity, which is what fascinates me. You need him to explain it." Compared to the theory of natural selection, religion came to seem silly, Rasmussen says, although her mother still hopes she will return to the fold.

Rasmussen entered the job market in the early nineties, just as molecular biology was beginning to emerge as a new way of doing avian systematics — the naming, description, and classification of birds. Morphology (the study of form and structure) was being superseded by molecular genetics, but Rasmussen wanted to study birds, not bird code. Owing in part to the peculiarly pre-Darwinian circumstances of her education, she wanted a job that would let her do what the nineteenth-century ornithologists had done: travel widely, observe and collect birds, and work with specimens in museums. In 1992, Rasmussen found that job, as the assistant to S. Dillon Ripley, the former secretary of the Smithsonian. Ripley's maternal great-grandfather, Sidney Dillon, had been the chairman of the Union Pacific Railroad, and young Dillon cultivated his interest in birds on the family's estate, in Litchfield, Connecticut. When he was a young man, he made several expeditions to India, in the grand style, meeting dignitaries while collecting birds. As a professional ornithologist, he wrote a ten-volume tome on the birds of India with the renowned Indian ornithologist Salim Ali. The Indian subcontinent, because of its turbulent political history and localized record-keeping, was one of the last regions on Earth where the avifauna was still not fully catalogued. Now, as his final project, Ripley wanted to produce a field guide that would be the last word on the subject.

In determining which birds to include on the final species list, Ripley was strongly predisposed in favor of museum specimens, and against photographs and eyewitness accounts, which would be allowed only if the evidence was what he deemed "diagnostic" — clear images, accompanied, if possible, by recorded birdsong. For Rasmussen, this meant examining tens of thousands of the 230,000 specimens from the region, which are scattered around museums in the United States, Great Britain, and India, as well as in Paris and Berlin. There would also be lots of fieldwork to do. "The job seemed like it was too good to be true," she told me. Not long after the project started, Ripley became too ill to continue working, and Rasmussen took over.

Like all branches of natural history, the traditional science of ornithology is based on specimen collections. The taxonomy of birds (the naming of species and subspecies) and their phylogeny (the ordering of species into ancestral families) were done using so-called study skins. With the recent advent of DNA-based research, scientists are now revising this scheme, working on a grand new system of classification, called "genetic phylogeny," which will reflect the new information gathered from DNA. What role specimens will play in the new molecular world is unclear, and some researchers, Rasmussen among them, fear that once DNA has been harvested from old specimens, future generations of ornithologists will see the skins as mere curiosities, artifacts of an earlier age.

Study skins differ from the mounted exhibits one sees in museums. Mounting birds, often in natural poses, is a good way to express what birders call the "jizz" — the bird's wild essence. But mounted birds are difficult to measure precisely, and minute variations in size are crucial in making taxonomic distinctions. In a study skin, the soft parts of the bird — brain and organs — have been removed, without disturbing the shape, so that the specimen resembles a dead bird in hand. If the skin has been skillfully dressed in the field, the feathers remain attached. The three largest collections are in London, New York, and Washington; among them, the natural history museums in these cities have almost 3 million skins. Most were gathered in the nineteenth and early twentieth centuries, before bird science developed professional degrees and standards. Collectors were often wealthy gentlemen who pursued ornithology as a hobby. Sometimes a museum would sponsor their birding expeditions to far-flung places, and in return the collectors would promise to contribute their skins to the museum. More sedentary collectors could also acquire skins from natural history dealers, such as Maison Verreaux, in Paris, or John Bell's shop, on Worth Street and Broadway in Manhattan, where as a young man Teddy Roosevelt bought specimens. Another popular source was the "plume hunters" who worked for the millinery industry, finding birds to provide feathers for ladies' hats. One of the most successful plume hunters of the nineteenth century, Joseph H. Batty, was also a collector for the American Museum of Natural History.

The Age of Collection waned in the twentieth century, mainly as a result of the growing popularity of the conservation movement. In his history of American ornithology, *A Passion for Birds,* Mark V. Barrow, Jr., explains that around the time the passenger pigeon, a once common North American bird, became extinct (the last one died in 1914, in the Cincinnati Zoo), the public began to grow aware of the possibility of species extinction and of the need to preserve birds in the wild. The nature writer John Burroughs called skin and egg collectors "men who plunder nests and murder their owners" and suggested that the "skin-collector should be put down, either by legislation or with dogs." He also called museum-based ornithologists "not only the enemy of birds, but the enemy of all who would know them rightly." Out of these sentiments grew modern birding, a phenomenally popular outdoor recreational activity (today, 46 million Americans call themselves birders) in which many collect not skins but "sight records" — eyewitness accounts, supported when possible by sound recordings and photographs.

Scientists regarded sight records as unreliable ("I wouldn't have seen it if I hadn't believed it" is how ornithologists refer to some birders' eyewitness accounts) and held fast to the principle that specimens were the ultimate standard in ornithology. (A recent example of skepticism about sight records is the controversy over the ivory-billed woodpecker, an extremely rare species that several birders claim to have observed and recorded in the southeastern United States, but which other ornithologists continue to maintain is extinct.) However, the ethics of conservation gradually won out over the interests of science, and legislation was passed in many countries that made collecting more and more difficult. This forced the scientists into a complex and paradoxical relationship with amateurs. On the one hand, they were at pains to distinguish themselves, as professionals, from the birders, whose standards they generally deplored. But on the other hand, because they were often unable to work with fresh specimens, the scientists had to rely more than ever on the skins amassed by the old amateurs. But who were they? What were their standards, these gentlemen scientists who did the early collecting and classifying?

Richard Meinertzhagen was one of the greatest of the old amateur collectors. His collection of 25,000 skins, most of which he shot himself, was acquired by Britain's Natural History Museum in 1954 — one of the largest private collections garnered by that museum in the twentieth century. In an obituary, the *Times* of London said that Meinertzhagen "will be remembered as an eminent and outspoken ornithologist of international fame and as one of the best and most colorful intelligence officers the Army ever had." The list of his military accomplishments would have been familiar to many readers: if not exactly a historic figure, he had been witness to a lot of history, especially in British East Africa and Palestine, where he served with T. E. Lawrence during the First World War. In later life, he made his reputation as the ultimate ornithologist, a sort of birding superhero. He could get to the remotest part of Afghanistan, find a bird no one had ever seen there before, shoot it, and prepare the skin in the field. (Toward the end of his collecting career, he managed to get around restrictions on killing birds by carrying a small gun concealed inside a walking stick.) Meinertzhagen always had beautiful skins, which is one reason that his collection was so valuable to scientists.

Rasmussen had never heard of Meinertzhagen when she began the Ripley Guide, but she soon came to recognize his name on many important specimens from the region. He was the sole authority for fourteen species and subspecies on the subcontinent. Like other noted British collectors who were posted to remote corners of the empire, he had shot and skinned and classified birds as a break from colonizing. (Another major collector in the region, Louis Mandelli, had been a tea plantation manager in Darjeeling.) Ripley had relied heavily on Meinertzhagen's skins in his work on the birds of India, and he had never said anything to Rasmussen about suspect specimens. ("He didn't talk to me much at all," Rasmussen says.) Salim Ali, Ripley's coauthor and a towering twentieth-century ornithologist, wrote favorably about Meinertzhagen in his autobiography, *The Fall of a Sparrow.* Rasmussen therefore had no reason to think, as she put it to me, "that there was anything wrong with the skins."

In 1996, when she was already a few years into planning the Ripley Guide, Rasmussen came across an article in an issue of *Ibis,* the journal of the British Ornithologists' Union. It was written by an Irish ornithologist named Alan Knox, a former curator at the British Natural History Museum who is now at the University of Aberdeen. Knox argued, convincingly, that several of Meinertzhagen's redpoll skins had been relabeled with incorrect data. Knox based his argument on the fact that skin collectors have characteristic styles of "making" a skin. Some slice off a small piece of the skull, in order to scoop out the brains, whereas others cut off the whole back of the skull, while still others take the brains out through the palate. Some birds are made with a full belly, others with a flat belly. The kind of thread, cotton, and internal supports used in making the skin can also differ from maker to maker. Based on an analysis of several preparers' styles, Knox concluded that at least two redpoll skins which Meinertzhagen claimed to have shot in Blois, France, on January 17, 1953, were probably stolen from a series of birds in the Natural History Museum, which had been collected by Richard Bowdler Sharpe decades earlier, in Hanwell, Middlesex, on November 17, 1884. Meinertzhagen had replaced the tags on the birds' feet with new tags, containing false data about where and when they had been collected.

On reading Knox's paper, Rasmussen was immediately concerned. "I thought to myself, if he went to the trouble to steal and change the data on a common bird like a redpoll, wouldn't he also try to fake some of the rarer birds in his collection?"

Several weeks later, Rasmussen traveled to Britain to meet Robert Prys-Jones, the head of the Natural History Museum's Bird Group. The museum's birds are kept in Tring, an hour's drive northwest of London, on the grounds of a private museum established in 1889 for Lord Walter Rothschild, an heir to the banking fortune and an insatiable bird collector. The collection is housed in a new building that abuts the museum, which is now open to the public. The bird skins, the feathers still as soft as the day they were shot, in some cases 150 years ago, lie on their sides in acid-free cardboard boxes, on tightly spaced sliding trays, in large white cabinets that form long spooky corridors stretching the width of the building. The corridors are filled with the intensely musty odor of moth cakes. Many of the older specimens bear traces of "arsenic soap," which collectors used to preserve their skins from insects. Among the historical curiosities in the cabinets are birds that James Cook brought back from the South Pacific and the finches collected by Darwin, among others, on his *Beagle* voyage, on the beaks of which the theory of natural selection was built. In the basement there is a long wall of cabinets filled with Lord Rothschild's beloved cassowaries, large flightless birds from New Guinea and Australia that used to roam freely in the park; now they're stuffed and mounted. (The vast majority of Rothschild's birds aren't in Tring, however — a subject that still causes pain within the Bird Group. Some 280,000 Rothschild skins were bought by the Whitney family for the American Museum of Natural History in 1931. Rothschild, sixty-three at the time, sold the birds for $225,000, in order to help pay off a peeress who was blackmailing him over an old affair by threatening to tell his mother.)

Rasmussen had been hoping that Knox was wrong about the redpoll frauds, but Prys-Jones assured her that he was not. After the *Ibis* paper, the British Ornithologists' Union asked Prys-Jones and Nigel Collar, an ornithologist at the conservation group BirdLife International, to evaluate the Meinertzhagen specimens. Prys-Jones had a Meinertzhagen redpoll skin X-rayed, and compared the image with an X-ray of a Sharpe skin. "They were obviously from the same maker," he told me.

The possibility of widespread fraud put Prys-Jones in a tricky position. His first responsibility, as head curator of the collection, was to the collection itself. He had heard rumors that the redpolls were not an isolated case. If fraud was pandemic, it might damage the collection at a time when some scientists were beginning to debate the value of keeping large collections. In the same way that card catalogues in libraries were disposed of once their contents were digitally rendered, so, perhaps, could specimens be removed from museums once they had been digitally sampled and photographed — freeing up valuable space for revenue-generating attractions like planetariums. "Serious people have seriously suggested that once you digitize the specimens you don't really need them," Rasmussen told me indignantly. "People are asking what collections are good for, why do we need to keep them?"

Nevertheless, Prys-Jones told Rasmussen that he would collaborate with her in trying to track down all the Meinertzhagen frauds among South Asian birds; together they would publish their results. "In the end, what else could we do?" Prys-Jones said.

Meinertzhagen was born in 1878, when the British Empire was at its zenith, and he lived to see its nadir. His mother, Georgina Potter, came from a well-known family of liberal intellectuals; her sister was Beatrice Webb, the Fabian socialist and cofounder, with her husband, Sidney, of the London School of Economics. The Meinertzhagen house was full of important late-Victorian thinkers and liberal politicians. As a boy, the biographer Mark Cocker notes, Meinertzhagen was taken on natural history outings by Herbert

Spencer, the philosopher who coined the phrase "survival of the fittest," and Thomas Huxley introduced him to Darwin, who allowed him to sit on his knee. Richard's father, Daniel Meinertzhagen, was a banker of German origin, who had little interest in his wife's liberal ideas and friends. He once fled the house when George Bernard Shaw came to visit.

Richard, called Dick, was taught to skin birds by Richard Bowdler Sharpe, whose redpolls he would later steal. Sharpe was one of those astoundingly energetic Victorians who collected and catalogued a vast number of natural history specimens of all types. He was the first curator of the Bird Room at the new British Museum (Natural History), which occupied Alfred Waterhouse's cathedral-like, terra-cotta-sheathed building that opened in 1881 on Cromwell Road, in South Kensington, to house specimens that had overflowed the confines of the British Museum, in Bloomsbury. (The name of the new building was officially changed to the Natural History Museum in 1992.) As a teenager, Meinertzhagen spent many hours skinning and classifying birds there. He also began keeping a diary of his ornithological exploits; this lifelong project would become as critical to his legacy as his skins.

Meinertzhagen had hoped to pursue his natural history interests at university, but his father had no faith in his scientific ability and sent him to work in the family banking firm, in the City. Meinertzhagen quit after a year to join the Royal Fusiliers, sailing for India in 1899. Moving to East Africa, in 1900, he discovered an avocation for killing. "I have myself felt the magnetic power of the African climate drawing me lower and lower to the level of a savage," he wrote in his diary. He satisfied his bloodlust by slaughtering animals and, occasionally, men. "I was surprised by the ease with which a bayonet goes into a man's body," he wrote in an account of an attack on an East African tribe in 1902. He seemed determined to reject the liberal pacifist philosophy of his mother's side of the family, which he called "a foul, infectious disease." However, he didn't entirely forget his scientific ambitions while in Africa: he identified a previously unknown species, the giant forest hog. Named *Hylochoerus meinertzhageni,* the hog is in the British Natural History Museum's collection.

In 1917, Meinertzhagen served as an intelligence officer for Sir Edmund Allenby, the commander of the British forces in Palestine.

While he devised ways to defeat the Turkish Army, he also studied ornithology, using calibration instruments at antiaircraft gun stations to calculate the speed and altitude of birds. In Palestine, he came to know T. E. Lawrence, who memorably describes him in *Seven Pillars of Wisdom:* "A student of migrating birds drifted into soldiering . . . He was logical, an idealist of the deepest and so possessed by his convictions that he was willing to harness evil to the chariot of good. He was a strategist, a geographer, and a silent smiling masterful man; who took as blithe a pleasure in deceiving his enemy (or his friend) by some unscrupulous jest, as in spattering the brains of a cornered mob of Germans one by one with his African knob-kerri." Meinertzhagen asked Lawrence to remove this account from later editions, but it remained in the book.

Palestine was the scene of the Colonel's most celebrated wartime action — the Haversack Ruse, in which he duped Turkish forces into thinking that the British would continue their drive toward Jerusalem by attacking Gaza instead of Beersheba. He allowed forged secret documents purporting to be detailed plans for a Gaza offensive to fall into Turkish hands, by making it appear that he had lost the haversack containing them while he was wounded and under pursuit. Not only were the orders and the plans faked, but there was also an easily broken code that allowed the enemy to intercept fake messages. The Turkish generals apparently fell for the ruse, and the British achieved a complete surprise and victory at Beersheba, which helped lead to their conquest of Jerusalem a month later.

After the war Meinertzhagen was a member of the British military delegation to the Paris Peace Conference, where, in opposition to Lawrence and the Arabists, he aligned himself with the Zionists, vigorously backing Arthur Balfour's declaration pledging British support for a national homeland for Jews (which was expressed in a 1917 letter to Lord Rothschild, who was a prominent Zionist). This was the beginning of Meinertzhagen's long support for the state of Israel (a square is named after him in Jerusalem). In Paris, the Colonel also had several memorable encounters with Lawrence, which he recorded in his diary and later used in his book *Middle East Diary 1917–1956,* published in 1959, twenty-four years after Lawrence's death. One evening in the summer of 1919, Meinertzhagen wrote, a distraught Lawrence appeared at the door of his hotel room. He confessed that the book he was writing about his experiences in the war was not all true, that he "had been involved in a huge lie — 'imprisoned in a lie' was his expression," Meinertzhagen wrote. Lawrence then went on to reveal that when he had been captured by the Turkish governor of Deraa, in 1917, he had been sodomized by the governor and his servants. Meinertzhagen begged Lawrence to tell the truth ("I loathe fakes," he said). In a later entry, Meinertzhagen said of Lawrence, "He has a trick of inflating the truth so that one cannot tell which is basic fact and which is embellishment."

In the early1920s, Meinertzhagen was a chief political officer in Palestine and Syria and then held important positions in the Colonial Office in London. He retired from the army at the age of forty-seven, with his pension, resolving to devote himself to ornithology and to spend more time with his wife, Anne Constance Jackson, whom he married in 1921. (He wrote in his diary that he had been offered several attractive diplomatic posts in other countries, including the governorship of the Falkland Islands and ambassador to Japan, and though he would have liked either, he turned them down because Anne didn't want to leave Britain, "and her happiness counts far more than worldly advancement.") Supported in part by Anne, who came from a wealthy family, Meinertzhagen traveled widely in pursuit of birds, including one long trip during 1925 and 1926 to Sikkim and Ladakh, when he collected many of his fourteen unique records on the Indian subcontinent; he published a two-part paper on his work in *Ibis*.

Three months after the birth of the couple's third child, Randle, in 1928, Anne, then thirty-nine, was killed in an accident on her family's estate in Scotland. She and Meinertzhagen were practicing outdoors with a revolver, and as Meinertzhagen was going to retrieve a target, Anne accidentally shot herself in the head while inspecting the gun. The death certificate said that she died of "injury to spinal cord & lower part of brain from a bullet wound at short range." She left the income from her estate to her husband on the provision that he never remarry; he never did. Meinertzhagen's cousin Theresa Clay, who was seventeen at the time of Anne's death, and thirty-three years younger than Meinertzhagen, became his protégée and ultimately his lifelong companion. They lived next door to each other in London, at No. 17 and No. 18 Kensington Park Gardens, but an indoor passage connected the two houses. Under Meinertzhagen's guidance, Clay became an expert in Mallophaga, a type of chewing louse that infects birds. The pair spent the 1930s traveling together. He would ask the curators of the great skin collections around the world, many of whom were his friends, whether he and Clay could use the collection in order to gather Mallophaga and further her work. In 1939 they spent one Sunday entirely unsupervised in the American Museum of Natural History, in New York, "shaking out" skins. Clay also accompanied Meinertzhagen on epic birding trips, collecting many birds herself. In 1937 the Colonel named his prize possession, the Afghan snowfinch, a species new to science, after her — *Pyrgilauda theresae.*

In his later years, Meinertzhagen received a series of honors, including Commander of the British Empire. He published a major work of ornithology, *Birds of Arabia,* in 1954, and in his final decade produced four volumes of memoirs, mined from his diaries, which became important primary sources for historians of British East Africa and of the First World War. He was bedridden during his last years, after a collision with an overenthusiastic dog in Kensington Park Gardens left him with a broken hip.

When Rasmussen and Prys-Jones embarked on the evaluation of Meinertzhagen's rare South Asian birds, they began with the assumption that the data on the tags were not reliable. They suspected that, as with the redpolls, Meinertzhagen had not only stolen the birds from a museum or a private collection but also invented the data. For this reason, Rasmussen feared that it would be impossible to determine who had actually collected the birds. But the answers were hidden in the skins. Rasmussen and Prys-Jones figured out ways to extend Alan Knox's basic forensic method, building a diagnostic file of the characteristic preparation styles of the major collectors in the region. In hundreds of cases, they X-rayed skins to reveal characteristics that weren't superficially obvious. Once they had identified a suspect Meinertzhagen skin as being made in another collector's style, they could check the museum's register to see if one of that collector's birds was missing. Often the bird that the Colonel had stolen would prove to be the best specimen of a series of birds shot around the same time. This method allowed Rasmussen and Prys-Jones not only to identify whom Meinertzhagen had stolen the bird from but also, by placing it within a series, to restore much of the data that Meinertzhagen had destroyed when he removed the label.

In several cases, they discovered that stolen birds had been remade by Meinertzhagen, probably in order to conceal the hand of the original maker. The most dramatic example of this type of fraud was *Athene blewitti,* a forest owlet, one of the rarest birds in India. The last reported sighting of it was made by Meinertzhagen, who claimed to have shot the bird at Mandvi, on the Tapti River in Gujurat, in October 1914. The previous forest owlet sighting was recorded in 1884, when James Davidson, a colonial official stationed in Western Khandesh, shot several in the jungle below the Satpuras, in Maharastra, about 300 miles away. Rasmussen and Nigel Collar, of BirdLife, who had once regarded Meinertzhagen as his hero, asked Prys-Jones to lend two of the Natural History Museum's owlets to the Smithsonian — the Meinertzhagen bird and one of the Davidson specimens. The Smithsonian's bird collection manager, Phillip Angle, slit the wing of the Meinertzhagen owlet, and along the humerus he found a yellowed piece of cotton that the Colonel had probably neglected to remove when he remade the bird. Rasmussen sent a sample of that cotton, along with a sample of cotton from the Davidson bird, to an FBI lab for fiber analysis. The fibers proved to be identical.

Eventually Rasmussen and Prys-Jones were able to prove that all fourteen of Meinertzhagen's unique records for species and subspecies in the Indian subcontinent were frauds. The Siberian accentor he claimed to have collected in Ladakh in April 1925 had actually been collected in Peking by a man named Kibort in September 1878. The three male willow warblers he claimed to have taken on the same day in January 1952 in Nagaland, northeastern India, had actually been collected between May and July of 1885 in northwest Siberia by Seebohm and Brown. His single specimen of Père David's snowfinch, which he said he collected in Sikkim in March 1952, in fact had come from Mongolia and was collected by Severtzov in 1877. Many more rare birds in the collection turned out to be frauds too. Meinertzhagen's two Blyth's kingfishers, which he reported shooting in Burma, actually came from the island of Hainan, in China. The Colonel's Andaman tree pie and his extremely rare Seychelles paradise flycatcher were also stolen and fraudulently labeled. In many cases, the false Meinertzhagen data had been incorporated into the subsequent literature; Ripley himself had used the false data in his books. Meinertzhagen brazenly published many of his false records in *Ibis,* "almost like he was daring people to catch him," Rasmussen said.

The profound unreliability of the Meinertzhagen specimens not only put the Ripley Guide in peril, it also threatened to undermine Rasmussen's faith in the efficacy of skins in general, on which she had based her career. Nothing else this large has been known to have occurred in the world of ornithology. The only case that comes close is that of the Hastings Rarities. Between 1892 and 1930, thirty-two of forty-nine new taxa added to the British List — the complete record of the more than five hundred species in the British Isles — were taken from within 20 miles of the town of Hastings, in southeastern England. In fact the birds were brought from other parts of the world and sold as genuine British birds to collectors eager to have a complete, up-to-date set. The chief suspect was a local taxidermist and gunmaker named George Bristow. When the frauds were finally detected, in the 1960s, some thirty species had to be removed from the British List.

But Meinertzhagen wasn't interested in money, as Bristow seems to have been. He wanted reputation. In this aspect, his actions resemble those of a British botanist, Professor John Heslop Harrison, of Newcastle University, in the1940s. Harrison was growing Arctic alpine plants in his garden, among other places, stealthily replanting them on the Isle of Rum, in the Inner Hebrides, and then using his "discoveries" as evidence to bolster his theory that vegetation predating the last Ice Age survived in some parts of the British Isles. He too was suspected during his lifetime, by an amateur botanist named John Raven, but Raven was never able to get a full accounting of the hoax published in a scientific journal.

To Rasmussen, one of the most shocking aspects of the case was the revelation that the authorities at the British Natural History Museum had known for years what the Colonel was up to. In 1995, Prys-Jones began to read the museum's files on Meinertzhagen, and he and Rasmussen discovered official memorandums showing that Meinertzhagen had been caught stealing birds as far back as August 1919, when, shortly after his return from the Paris Peace Conference, he was apprehended with nine birds in his briefcase as he left the museum. He was subsequently banned from the Bird Room, but managed to get reinstated in 1921, with the help of Lord Rothschild, even though he too suspected Meinertzhagen of stealing birds from him. The Colonel had been caught stealing skins again in 1934, and in 1935 he was investigated by Scotland Yard for having removed a volume of the journal *Parasitology* from the museum's library and for tearing pages out of another volume. But charges were never pressed, and Meinertzhagen escaped public censure. On the contrary, he was made an honorary associate of the museum in 1954 and was issued a pass key.

Prys-Jones thinks that some individuals at the museum might have been afraid to accuse Meinertzhagen out of fear of losing his skins. In 1947, Meinertzhagen had threatened to bequeath his collection to the American Museum of Natural History. The eminent ornithologist Ernst Mayr, who was then the curator of the museum's Rothschild collection, called the Meinertzhagen skins "the finest private collection of birds in existence" and confessed that he was "almost breathless" at the prospect of acquiring it. But the British birdmen, having lost the Rothschild collection to New York twenty years earlier, weren't about to lose the Meinertzhagen specimens. "I think some individuals at the museum might have figured, if he stole birds from them, it didn't matter so much, because they'd get the birds back in the end, when he left them his collection," Prys-Jones told me. "I don't think they knew he was changing the tags, which is a far more serious matter." In any case, the interests of the gentlemen's club prevailed over the interests of science. As was the case in the celebrated fraud involving the Piltdown Man — the fossil, "discovered" in Sussex by Charles Dawson and unveiled in 1912, that purported to show a missing link between the Neanderthals and the Cro-Magnons — the clubbiness of science would keep the truth buried for decades.

As Rasmussen and others began to publish papers on the frauds, she met people who, in one capacity or another, had known Richard Meinertzhagen. A sixty-eight-year-old retired librarian from Tring, Effie Warr, remembered Meinertzhagen from the time she had spent working in the Bird Room in London when she was a young woman. "The museum had given him his own gallery," Warr told me, "and that was where he kept his birds. He was quite old by that point, and nearly deaf, so you really couldn't talk to him, other than say hello. We were told to keep an eye on him, but they never accused him of anything, because they didn't want it to get into the public domain. Because they wanted his birds, you see."

It turned out too that in his great work, *Birds of Arabia,* Meinertzhagen had relied upon the unpublished manuscript of another ornithologist, George Bates, who had died before finishing his own book. Meinertzhagen had used a copy that he had taken from the British Legation offices in Jidda, not realizing that another copy was held in the British Natural History Museum. The duplicate copy, Warr said, revealed the extent of Meinertzhagen's borrowings.

Rasmussen also heard from some of the Colonel's old friends, who continued to support him. In a review of a book that included an article by Rasmussen and Prys-Jones about Meinertzhagen, Bill Bourne, a senior British ornithologist, wrote that "the usual people dance on Colonel Meinertzhagen's grave," noting that he "did more good work than most of their critics put together."

As the ornithologists were exposing the Colonel's bird-skin frauds, several researchers were examining other aspects of his life and legacy. One of these was an independent T. E. Lawrence scholar, J. N. Lockman, who published *Meinertzhagen's Diary Ruse* in 1995. Meinertzhagen habitually typed his diary entries and left behind no handwritten notes, so it is impossible to know to what extent he rewrote or embellished his life and observations. Lockman concluded that "Meinertzhagen's twelve Lawrence entries are virtually complete forgeries of much later date, cleverly conceived but sloppily executed." Diary entries about Lawrence, such as "I believe I was the only one of [Lawrence's] friends to whom he confided that he was a complete fraud," dated 1919, were actually written in the late fifties. (Seen in this light, Meinertzhagen's account of Lawrence saying to him, "Someday I shall be found out," may really be Meinertzhagen talking to himself.) Lockman pointed out that the suspect diary entries that he saw were typed on a different kind of paper from the surrounding entries, with a different typewriter ribbon, and he also noted that the sequence of page numbers does not match the pages immediately preceding and following the entries.

In 2001, Rasmussen met a Los Angeles writer named Brian

Garfield, the author of *Hopscotch* and *Death Wish,* among other novels, who was working on a nonfiction book about Meinertzhagen. Among the mysteries Garfield was researching was an old rumor, apparently widespread at the time, that Anne Jackson's death wasn't an accident. Garfield suspected that Anne, who was also an accomplished ornithologist (she was an honorary member of the British Ornithologists' Union), had discovered what her husband was up to. She had accompanied him on portions of his long South Asian birding trip in 1925 and 1926, and she might have known that many of the rare birds he claimed from this trip, published in his *Ibis* paper, were not in fact his. When she threatened to expose him, Garfield theorized, the Colonel flew into a rage and shot her.

Garfield, whose book *Meinertzhagen Mystery* will be published this fall, wrote to me recently, "I can add only that my own personal belief — not a fact but an opinion — is that the two doctors' description of the bullet path suggests an angle of entry and exit that is not consistent with accidental discharge of a revolver while it is being checked. To me it suggests that the handgun was slightly above her head, pointed at a downward angle, so that the bullet would enter the frontal lobe and exit farther down through the spinal cord. (If you were to shoot yourself in the eye while looking down the barrel . . . you'd blow off the top or the back of your head — not your spinal cord.) Annie was vigorous but not very tall. Richard was about six feet five. Nobody else was present. No inquest was held. No one was charged with any crime."

Meinertzhagen deposited the seventy-six leather-bound volumes of his diary with the Bodleian Library of Commonwealth and African Studies at Rhodes House, at Oxford University — some 4 million words chronicling his life as a soldier, birder, spy, and observer of historical personages. Randle Meinertzhagen, a seventy-eight-year-old retired investment banker who is the Colonel's only surviving child, controls access to the diaries. Ran, as he is known, bears a striking resemblance to the Colonel. He seems like a decent man in the difficult position of trying to protect his father from charges that he knows are at least partly true. "I think the old rogue did take a few birds," he told me when we met in London, but added that he probably had some good reason for doing it. "He believed the birds weren't properly skinned or cared for and he could do a better job of it himself." For his father, the ends always justified the means. Ran had a rather distant relationship with him. "If I ever asked him about himself," he told me, "my father would say, 'You can find out anything you want to know about me in my diaries,'" which were kept in the drawing room in No. 17 Kensington Park Gardens. Meinertzhagen would add, "Just don't ask me anything about them afterwards."

I spent several days in Rhodes House, reading the diaries. I saw an entry from 1913, when Meinertzhagen returned to birding after years of pursuing larger game: "I fell to the altar of ornithology, as the distressed brain takes refuge in religion." In 1924, on hearing of a report that George Bristow, the Hastings taxidermist, was suspected of claiming birds killed elsewhere in Europe as British birds, he writes, "I know of several cases of disloyalty among egg-collectors, but this is the first occasion on which I have come across dishonesty among skin collectors." His only comment on his wife's death, in 1928, is "We had been practicing with my revolver and had just finished when I went to bring back the target. I heard a shot behind me and saw my darling fall with a bullet through her head." However, in the year following her death there are several long, anguished descriptions of a sort of soul sickness, among them: "My life has been transformed into a desert, which no persons can now people."

In addition to the diary entries, there are many photographs of the beautiful Theresa Clay. She is sometimes depicted in the act of "shaking out" a bird skin for Mallophaga. There is also one nude photo, taken at the beach when Clay was around the age of fifteen.

In 1948, while collecting in the Himalayas, Meinertzhagen writes of receiving a letter from Clay in which she expresses her desire for independence, saying (he reports), "I have to face up to the fact that I shall be left alone at a time when a woman is most likely to feel lonely." Meinertzhagen writes, "Life for me without her would be unbearable, unthinkable, and utterly impossible," and he goes on, "I feel like some caterpillar who for well nigh twenty years has been browsing on a beautiful shrub and has stripped it of leaves and flowers." In the end, however, Clay did not leave him; she remained his companion until his death, marrying only when she was in her mid-sixties. She died, an esteemed entomologist, in 1995. The chewing lice that she and Meinertzhagen collected also went to Britain's Natural History Museum.

Historical artifacts of many kinds are pasted into the diary. One is a letter from Chaim Weizmann, dated June 14, 1948, shortly after Weizmann became Israel's first president, in which he tells Meinertzhagen that he is "deeply grateful . . . to you for your friendship and confidence and help through all these years." There are three pigeon feathers with coded messages written on the shafts. There is careful documentation of the Haversack Ruse, including a letter from the wife of the haversack's owner announcing the birth of a son, which Meinertzhagen forged to add a personal touch to the contents, and a handwritten letter, in Turkish, supposedly found in the rubble of Gaza, that recounts in credulous terms the interception of the Colonel's haversack.

But if there is a Rosebud in the diaries — a key that unlocks Meinertzhagen's motives — I didn't find it. I came away with the feeling that the pages were like the skins: some are no doubt authentic, but so many others may contain false data that one can't be sure what to believe. The Lawrence passages are obvious forgeries, for all the reasons that Lockman pointed out in his book, but as he noted, these were perpetrated when the Colonel was nearly eighty and more careless: "Only as an eighty-year-old could the once master forger have so lost his touch." What about a memorandum that Meinertzhagen supposedly wrote to David Lloyd George, the British prime minister, dated Paris 1919, which predicts with uncanny accuracy the 1948 Arab-Israeli war? Could this have been written after 1948? What about the pigeon feathers? The Weizmann letter? It's not even clear that Meinertzhagen conceived the Haversack Ruse or that it actually fooled the Turks. But if the ruse is itself a ruse, then which is the better ruse? And where does the ruse end?

From time to time during her research, as a welcome break from specimen work, Rasmussen went birding on the subcontinent. She made expeditions to India (where, near Darjeeling, she was walking through a rain forest under an umbrella and realized that the fat drops slithering down the sides were actually leeches), the Andaman Islands, Burma, and the Himalayas, in many cases visiting the places where Meinertzhagen had collected his rare birds.

Rasmussen knows how to shoot birds (she learned in Patagonia, where she collected cormorants) and also how to skin them, but she didn't use her skills very often on these birding trips. (However, she did manage to collect a few specimens in Burma, using a fine-mesh net to entrap the birds. "Then you just give them a little squeeze and that's it, they're dead," Rasmussen said tenderly, adding, "I mean, think of how many birds die because a skyscraper goes up in a migratory area.") But almost everywhere she birded on the subcontinent, taking specimens was politically impossible. "Now, with avian flu, it's even worse," she said.

Sometimes, in the course of her research, she found herself in awe of the magnitude of the Colonel's fraud. He was, for better or worse, the last man to unite science and birding, before the great schism. (His rarest bird, the *Pyrgilauda theresae,* turns out to be authentic.) Occasionally, she said, she had the feeling that Meinertzhagen was her father, although, being of a scientific cast of mind, she doesn't put much stock in notions like this. "In some ways, Meinertzhagen does remind me of my father," she told me. "He was brave, tall, strong, he liked killing animals, and he was unreliable. But that's pretty much where it ends."

The most memorable of her birding trips took place in 1997, when Rasmussen went to India in an unlikely attempt to find *Athene blewitti,* the forest owlet. Without the Meinertzhagen specimen to skew the data, Rasmussen guessed that the bird's range might be much narrower than previously thought, and that ornithologists had been looking for the bird in the wrong places. She searched the jungle near where Davidson, from whom Meinertzhagen had stolen the owlet, had collected his birds. But most of the jungle had been cut down, and after several days of looking, Rasmussen was ready to admit that the owlet, lost to science for 113 years, was truly extinct.

On the day before the expedition was due to end, Rasmussen began as she often did, at 4 A.M., listening for owls. By eight- thirty it was getting hot, and Rasmussen was opening her water bottle when the expedition's leader, Ben King, said, "Look at that owlet," pointing up into a tree. Rasmussen looked and said in excitement, "It doesn't have any spotting on the crown and the mantle!" (The common spotted owlet displays such features.) The forest owlet remained there long enough for Rasmussen to shoot it. The video was diagnostic, and the owlet made the list.

The birds of South Asia represent only a fraction of the 25,000 birds in the Meinertzhagen collection: the European and African skins, which could be equally riddled with frauds, await examination by other researchers. Prys-Jones thinks that, ultimately, 5,000 of the skins in the Meinertzhagen collection could turn out to be fraudulent. Rasmussen and Prys-Jones are preparing a paper which will enumerate other frauds in the South Asia birds.

Rasmussen, who is now an assistant professor of zoology at Michigan State University, says that while the Meinertzhagen frauds tested her belief in the reliability of specimens, ultimately her detective work with Prys-Jones and others reaffirmed the importance of the skins. "If this fraud had been done only with photographs, or with sight records, we never would have been able to figure out what happened," she told me.

In addition to the bogus Meinertzhagen birds, Rasmussen removed eighty-five other species from the final South Asia list, categorizing them as "hypothetical," largely because she judged the sight records supporting them to be insubstantial. Her decisions were praised by the scientific journal *Nature*, which, in a glowing review of the Ripley Guide, called Rasmussen "brave" and said, "The book's greatest value is that Rasmussen has taken nothing for granted, even information published in Ripley's own works." But the reviewer for a leading birding journal, *British Birds*, chided Rasmussen and her lead illustrator, John Anderton, for excluding so many "undoubtedly valid recent field observations." He added, "Doubtless this reflects the fact that the authors are primarily museum workers."

BILL SHERWONIT

In the Company of Bears

FROM *Anchorage Press*

I USED TO HAVE nightmares about bears. They entered my dreams in the 1970s, shortly after I came to Alaska, and roamed the forests of my subconscious for many years after.

A geologist just out of graduate school, I spent my first Alaska summers in some of the state's wildest, most remote grizzly bear country. Each year, usually toward the end of the field season, phantom grizzlies would stalk me, chase me, attack me. They lurked in my dream shadows, ominous and haunting.

I now sometimes wonder if those nightmares were omens. Perhaps they spoke of things to come, of a summer afternoon in Shuyak Island State Park, at the northern end of Alaska's Kodiak Archipelago. . .

Five of us have spent the morning in the kayaks; now its time to stretch our muscles and explore one of the many small islands that border Shuyak's northern coast. The islet we choose is inhabited by Sitka black-tailed deer, and as we approach we see several feeding in open meadows. The islet is also home to a female brown bear with three tiny cubs. We spotted them earlier in the day, though the bear family has since disappeared into the forest.

That was my first sighting of brown bears, the costal cousins of grizzlies. Alaska's brown bears tend to be more chocolate in color and have less distinctive humps and shorter claws that the grizzlies of the interior do. On average they're much larger animals, mainly because they have access to more plentiful, energy-rich foods, especially salmon. A large male grizzly may weigh 600 to 700 pounds in fall, when it's fattened for hibernation. The largest brown bears are twice that size. And nowhere do brown bears grow larger than on the Kodiak Archipelago, home to the subspecies *Ursus arctos middendorffi*. Even here, researchers say, adult females rarely reach 700 pounds, though this mother bear appeared much bigger.

We beach our kayaks, then split up. I go with Sam, one of the expedition's guides, following a game trail that begins in meadow but soon borders a thick stand of spruce. Sam calls out to announce our presence: "Hooyah! . . . Hooyah!"

The trail peters out where the forest reaches the island's edge. We have a simple choice: return down the trail or cut through the woods. Sam chooses the trees. I follow, despite some misgivings. He's the guide, after all.

The spruce are 20 to 30 feet high, spindly, and densely packed. We can't easily see more than 10 to 15 feet ahead, sometimes less. We're walking slowly, talking loudly, when suddenly my worst nightmare comes true: a bear charges out of the forest's shadows.

She must have tried to hide her family in this stand to avoid the strange two-legged invaders of her island. But we've entered her sanctuary and, however innocently, threatened her offspring. Retreat hasn't worked, so her only option now is to defend her cubs by force.

Things begin to speed up and simultaneously move in slow motion around me. Less than 20 feet away, the bear is a blur of terrible speed, size, and power, a dark image of unstoppable rage. Her face is indistinct; I sense more than see her teeth and claws. Two giant bounds are all it takes for her to reach Sam, 5 feet in front of me. Somewhere, amid the roaring that fills my head, I hear a cry: "Oh no!" I'm certain that Sam is about to die, or be seriously mauled, and fear that I may be also.

The last thing I see is the bear engulfing Sam. Then, despite everything I've learned about bears, breaking one of the cardinal rules of bear encounters, I turn and run.

My instincts tell me to get out of sight, out of the woods. Climbing one of the slender trees isn't an option, and without a weapon, there's not much I can do to help Sam. The only question now is whether the bear will come after me when it's done with him.

I run out of the forest onto a narrow stretch of beach. I have to find the other three members of our party, get Sam's rifle from his kayak, and try to rescue him.

Back in the forest, Sam is doing what he must to survive. Later he tells us that as the bear charges, he ducks his head and falls backward. Falling, he sees the bear's open mouth, its teeth, its claws. Hitting the ground, he curls into a fetal position, to protect his head and vital organs and offer the bear a shoulder to chew on instead. With the bear breathing in his face, he plays dead.

The bear grabs Sam in a "hug," woofs at him, and bats him a few times. But she strikes him with her paw, not her claws. There's no sound of tearing flesh. And when, after several moments, there is no response from her victim, she ends her attack just as suddenly as she began. The treat removed, she leaves with her cubs.

I'm still standing on the beach, listening and looking for any sign of the bear, when I hear Sam shout, "The bear's gone . . . I'm all right." Miraculously, he's uninjured except for a small scratch on the back of his hand that he got when he fell backward into a small spruce. For someone who's just been attacked by a bear, he's taking the incident calmly, much more calmly than I would. Later I learn that he'd had some experience. He'd been "false charged" by bears three times before.

"Thank goodness it was a friendly bear," Sam says. "It wasn't looking for a fight; it was trying to make a point: 'Leave me alone.'"

Heading across a meadow to warn the others, we see the sow 100 yards away, still agitated. She stands up, then falls back to all fours and runs around in circles and stands up again. She's looking down the island, and we guess she has seen or smelled our companions. The bear stands one final time, then turns sharply and lopes into another spruce stand. Strung out in a line, the three cubs run hard to keep up with Mom.

We rendezvous with the others and leave the island. Back in camp, we talk for hours about the encounter and second-guess ourselves. We agree it was foolish to visit the island, given our earlier bear sighting, and even more foolhardy to cut through the woods.

I have never carried a gun into Alaska's backcountry. I'm not a firearms expert, have no desire to be one, and believe that guns cause more trouble than bears. Like Sam, I also believe that guns change the way a person relates to wild places and wild creatures. They offer security, but they also can prompt people to take chances they ordinarily wouldn't, sometimes resulting in confrontations that might have been avoided. The usual result is injury or death, often for the bear.

For a while after the Shuyak attack, I questioned my philosophy. It's often said that bears are individuals. Each is different and unpredictable. The writer and naturalist Richard Nelson, in *The Island Within,* says, "All it takes is once in a lifetime, the wrong bear in the wrong place. Without a rifle (and the knowledge of when and how to use it), the rest of the story would be entirely up to the bear . . . [The rifle is] my way of self-preservation, as the hawk has its talons, the heron its piercing beak, the bear its claws." Yet as time has passed, I've become more convinced than ever that it's right for me to walk unarmed in Alaska's backcountry. It would be different, perhaps, if brown or black bears preyed on people. But they rarely do. My choice is a gesture of respect to the animal and its world. I'm only a visitor in the bear's realm.

On Shuyak, we provoked the attack. A mother was crowded, and she wanted to eliminate what she perceived as a very real threat. She was protecting her cubs, no more, no less. Playing dead, removing the threat, proved the best thing to do, not fighting back. Shooting her would have been a tragedy.

Bears still pad through my dreams, but they don't menace me anymore. I've come to believe that Shuyak bear was a messenger. Her attack was fiercer than anything I'd imagined or dreamed, yet it ended well for both bear and humans. In the years since, I've spent many days in the company of bears and have learned that they are remarkably tolerant of people — much more tolerant of us than we are of them.

Bears, especially the brown and grizzly, have been my guides as I find my own wild spirit.

It's a dozen years after the Shuyak encounter. I'm on the upper Alaska Peninsula, camping with several other bear lovers. We've barely finished storing food and setting up tents among tall meadow plants when a family of brown bears ambles down the narrow sand beach by our camp.

The bears are walking slowly, without menace, eating as they go. Still, they're moving ever closer. In the lead is a large female with a beautiful milk-chocolate-colored coat. Trailing her single file are four tiny cubs. Darker than their mom, the cubs were born in January or February. Only a pound or so at birth, the largest of the four now, in late July, weights 25 to 30 pounds. The runt weighs half as much; small and fragile-looking, he's like a real live teddy bear.

In most places the steady approach of a brown bear family would be cause for concern, perhaps even alarm, because females with cubs are notoriously aggressive. The five of us who watch the bears, soon to be joined by a half-dozen others, would likely announce our presence by talking loudly, maybe while waving our arms and slowly retreating. We would crowd together, to increase our size, and worry that the bears might invade our camp. Someone in the group might reach for a firearm or pepper spray. None of that happens here.

Newly arrived at Alaska's McNeil River State Game Sanctuary accompanied by a Fish and Game bear biologist, we watch calmly and quietly from the edge of camp while the mom and her cubs forage among sedges that grow where beach sand gives way to broad mudflats. We whisper of our good fortune as the bears come within 100 feet, then 80, then 40. It's clear their approach has nothing to do with us; it just so happens that the mother's beach foraging has brought the family in our direction. The cubs look toward us now and then with an apparent mix of curiosity and anxiety, but their mom hardly pays us any heed as she gulps down clumps of green blades.

This sow is one of dozens of brown bears that congregate at McNeil each summer. Larry Aumiller, the sanctuary manager, recognizes her: Snow Bear. She's known to be very tolerant of humans and comfortable in their presence. In her own ursine way, she understands that we mean no harm to her or her cubs.

McNeil's staff has named nearly all of the sanctuary's adult bears, at least the regulars who come back year after year. Some people have criticized the practice, but Aumiller, who's run the McNeil program since 1976 (and who retired in 2005, after thirty years at the sanctuary), says that with more than eighty adult bears, "you need a way to keep track of different ones. Naming them is almost like a mnemonic device . . . It doesn't imply human characteristics or qualities."

Snow Bear got her name several summers ago when she sat upon a patch of snow. Now in her prime, she weights 500 to 600 pounds. That's a good-sized female. McNeil's largest males weigh twice that.

Snow Bear is only one of the several females to bring new litters to McNeil River this summer, but she and her cubs dominate our conversations over the next four days. Partly that's because she frequently forages near camp, but it's also because four-cub litters are so rare. The usual number is two. Over the past quarter-century, only four females have shown up here with four cubs.

Brown bear cubs are usually weaned from their moms during their third summer, at age two and a half. The chance of a cub surviving that long at McNeil is fifty-fifty, says Aumiller. Some drown; others starve or are killed by disease, falls, or other bears.

The runt's odds are even slimmer. A few days later, during a swim across McNeil Lagoon, he lags behind, and for a few moments we wonder if he'll make it to shore. He does, but takes far longer to recover than his siblings.

The McNeil sanctuary, established in 1967 and managed by the state's Division of Wildlife Conservation, protects the world's largest gathering of brown bears. The focal point of this gathering is McNeil Falls, where bears come to feed on chum salmon. During the peak of the July-August chum run, dozens of brown bears congregate there. As many as 144 adults and cubs have been identified along McNeil River in a single season. In July 1997, biologists counted 70 bears at the falls at one time.

Perhaps even more impressive than the numbers of bears is their acceptance of a human presence. Every day from early July through late August, ten visitors plus sanctuary staff spend seven to eight hours at the falls while stationed at two gravel pads. "Thing about it," Aumiller says. "You've got this group of people standing in the middle of forty, fifty, sixty bears. You're very close to where they want to be. And they tolerate you." Some will eat salmon, take naps, nurse cubs, or even mate within a short distance of the viewing pads.

This high degree of tolerance may be Aumiller's greatest legacy. More than anyone lese, this sixty-something naturalist with dark, silver-speckled hair, sparking eyes, and an easy smile has demonstrated that bears can become habituated to people without also becoming food-conditioned. And he's shown that such bears — Aumiller calls them "neutrally habituated" — are safe to be around. As he explains it, "At McNeil, humans are neither a threat nor a source of food. Over time it became clear from their actions that the more tolerant bears were perceiving us as neutral objects, maybe as innocuous as a rock or a tree."

Before the McNeil experiment, most people, including many so-called bear experts, believed that habituated bears, particularly grizzlies, are extremely dangerous. McNeil proved that the opposite is true, if food is removed from the equation. For that reason, all human food is stowed away at McNeil, and meals are prepared and eaten in a sturdy cabin that's off-limits to bears.

Reflecting on the lessons he's learned in nearly three decades at the sanctuary, Aumiller readily admits that living with bears is not an easy thing, but McNeil is proof of what's possible when humans are willing to compromise: people and bears can peacefully coexist, often in close company.

"What goes on here is still news to a lot of people," he says. "They don't think it can happen. But it does. McNeil shows that if you learn about something that's different from you, and begin to appreciate it, then you'll figure out a way to keep it in your life. You'll learn to coexist."

The number of bears gathering at McNeil Falls is down this summer, reflecting a poor return of chums. Unless there's a sudden late surge of fish, the run is likely to be the smallest on record. With less food to go around, bears are less tolerant of each other as they compete for prime fishing spots. There's more tension, more fighting, and fewer scraps for the adolescents and the females with cubs that roam the river's perimeter. Large, mature males weighing up to 1,000 pounds or more and bearing numerous pink battle scars hog the best fishing spots.

With little hope this year at the falls, many adolescents and females with cubs have taken to wandering the costal mudflats, where they feed on sedges and other greens or chase salmon in tidal channels.

Once, as we're watching from a remarkably close distance, Snow Bear rolls over onto her back to nurse her cubs. With her in this most vulnerable position — legs extended and her more sparsely furred belly and groin area revealed — I can easily see, or at least imagine, a physical likeness between human and bear.

As David Rockwell explains in *Giving Voice to Bear,* many Native American tribes, including some Alaska Natives, have attached great meaning to the physical similarities of our two species. Bears, like people can easily stand on their hind legs. They can even walk upright for short distances. And when a bear is standing, the animal's front legs hang at his side like arms. As anyone who's seen bears catch and eat salmon knows, they are surprisingly adroit when using their front paws. Even more impressively, Rockwell notes, "in captivity they have been known to peel peaches." And there's this fact: skinned bears look eerily human.

Such resemblances, combined with similarities in diet, and in behaviors (strong maternal instincts, problem-solving abilities, playfulness), have led many Native groups to feel a special kinship with bears. Tribes throughout North America traditionally believe the bear to be half human, or that humans were descended from bears.

I like that notion. I've come to relish the shape-shifting myths that tell of bears becoming humans or humans becoming bears. Something about those stories resonates. They hold a certain magic. And here things start to get tricky: How do you talk or write about such things without seeming New Agey? Without trivializing something that holds great power? Or without stealing from other cultural traditions?

I don't know the answer, so I usually keep my more mystical leanings under wraps. But here I'll plunge ahead: Bears are one of my totem animals. Though I come from a different cultural tradition, this white guy of European descent, Lutheran upbringing, and scientific training nevertheless finds meaning in the Native American notion of totems.

A totem animal is one that has caught or demanded my attention, one with which I sense some special connection or have an inexplicable fascination. A teacher. A messenger. A guide. Bear is one. Squirrel, chickadee, wolf, and spider are others.

As weird as it seems, there's something to the kinship/totem idea, even for a recovering, born-and-bred fundamentalist Christian and ex-geologist. Bear brings me closer to myself.

I recall no special interest in bears during my boyhood years in Connecticut. Bears weren't part of my world, though it seems they must have occasionally roamed through the woodlands my friends and I explored. What I knew of them I learned from sensationalized outdoor-magazine stories and "bear tales" sort of books; namely, that bears are dangerous, unpredictable, bloodthirsty critters that can — and will, if given the chance — tear you limb from limb.

Not until I came to Alaska in 1974 did I see a bear in the wild. And it wasn't until the late 1980s — after I'd left Alaska, switched careers from geology to journalism, and then returned to work at the *Anchorage Times*, eventually to become the newspaper's outdoor writer — that I became a passionate student of *Ursus*.

My education accelerated dramatically, and my affinity for bears deepened immeasurably, during and after my first trip to McNeil, in 1988. I've been back a half-dozen times since. Taught by both the bears and Aumiller, I've been changed by McNeil, as so many others have been transformed. I've come to understand bears as complex, amazingly adaptive, and intelligent creatures.

As the sanctuary's "take-home" lessons took hold, my fear of bears gave way to respect and reverence. Yes, bears — especially grizzlies — are large, powerful, and sometimes fiercely aggressive animals, particularly when protecting their young or food. They have the strength and tools to do great harm. They can be dangerous. But they're not bloodthirsty killers of humans and rarely prey on people. And I've learned they carry a power that goes far beyond their physical strength and size and ferocity.

Shortly before eleven on my last evening at the sanctuary, I stand above the sand beach in gray light that is dimming to darkness. The land and seascape beyond our small campground is once again a place only of bears, gulls, eagles, ravens, ground squirrels, foxes, and salmon. Watching from camp's edge, I count twenty-nine bears scattered from the mouth of McNeil River to the outer mudflats of Kamishak Bay. And I hear the sounds of night: the screech of gulls and eagles, the loud splash of bears chasing fish, the melancholy three-note call of a golden-crowned sparrow in alder thickets behind camp.

The bears are distant silhouettes. I see one family with two cubs, another with three. And then I see Snow Bear, still with her four. This cheers me, even while knowing the odds are small that they'll all make it through the summer.

The wildness deepens with the darkness as we humans shrink into our tents and sleeping bags, hiding from the night, the cold, the rain. Thick clouds shroud the landscape as salmon move through tidal flats toward McNeil River and bears hunt the salmon as they've done for centuries.

Mid-August, the Brooks Range. Two years after meeting McNeil's brown bear quadruplets, I am alone, camped deep in the Arctic. For the past two weeks I have been keeping watch for bears. Partly that's to ensure I don't have any surprise encounters with a grizzly. Unlike their coastal relatives, grizzlies don't easily adapt to being around other bears. Or people. Researchers guess that's because they rarely, if ever, gather in large groups to feed on concentrated, energy-rich food sources. So they haven't evolved the "social behaviors" to be in close company with each other — or, by extension, with humans.

My pervious experience with grizzlies has been this: upon seeing, smelling, or otherwise sensing *Homo sapiens,* most skedaddle with little or no hesitation. They want nothing to do with us hairless bipeds. Still, if you surprise a bear at close range, all bets are off. The creature must make a split-second decision. It may flee, or it may choose to forcefully eliminate the perceived threat by attacking. Happily, surprise encounters are rare in the vast tundra expanses of Alaska's northernmost mountain chain.

There's another, deeper reason that I've been keeping watch: I wish to share the landscape with bears.

In one respect, I have been doing that since arriving in the Brooks Range. Whether or not they're physically present, grizzlies regularly pass through the valley I've been exploring. These "barren ground" grizzlies are almost constantly on the move in their search for food, which is not nearly as abundant here as it is along Alaska's southern coastline. For all its beauty, most of the Arctic is a place of scarcity — except for its mosquitoes. The living is not easy, especially for a large predator.

In the course of my travels, I've seen plenty of bear sign: tracks, scat, and tundra diggings, where a grizzly has hunted ground squirrels. But no bear. So I wonder: Will seeing a grizzly make my trip more memorable? Is it not enough to know that bears frequent these valleys and ridges? I know I won't be disappointed if I leave without seeing grizzly, but seeing a bear would inevitably enhance my stay.

Sitting in camp, I imagine a grizzly walking across the tundra. Sitting on a tundra bench. Watching. Watching me? Might bear seek me? Certainly I've come to feel a more ursine presence in my psyche in recent years.

I ask myself again why I've come to this Arctic wilderness and its high alpine. There's no question that adventure is part of the draw. So is the wild beauty of this place. Since I first came here more than a quarter-century ago, the Brooks Range has become my favorite wilderness. It's a place where wildness is manifested in wave after wave of rocky, knife-edged ridges that can be attained without climbing expertise, and which stretch to the horizon and beyond; in glacially carved basins that grow lush in midsummer with the rich greens of tundra meadows and the vibrant purples, yellows, magentas, and blues of alpine wildflowers; in wolves, Dall sheep, caribou, bears, and wolverines; in an unpeopled landscape where on can still travel for days, even weeks, without seeing another person, or even signs of humanity.

The Brooks Range is where I saw my first wild bear, in 1974, while working as an exploration geologist. A chocolate-colored grizzly, the animal stood several hundred yards away, busily digging in the tundra for roots or perhaps for ground squirrels, while I collected rock and sediment specimens from a mountain stream. I was deeply stirred by the presence of the bear, but what moved me initially was the grizzly itself rather than anything it implied about wildness.

More than adventure or memories is at play now. I want to understand better what wildness means to me. I want to embrace the wild other that roams this Arctic world and also lurks inside me.

Keep searching and wondering, I encourage myself. Sleep on it. Dream it. What's important? What's here for me? What's the essence of this journey? Stay open to possibilities, to miracles and revelations. Let the landscape speak. Pay attention to its inhabitants: animal, plant, rock, creek, mountain.

Three days later in the Brooks Range, bear comes to me.

This August day has been gray, raw, windy, and wet, so I've spent most of the morning and afternoon hunkered in my dry tent and warm sleeping bag. Finally braving the storm, I duck under a small tarp and savor a meal of granola bars, cheese, oatmeal, beef jerky, and pressed ground coffee. Then I scan the landscape.

Across the creek, on a gentle rise, I see a couple of dark, hulking shapes. That's nothing unusual; I've seen dozens of "bear boulders" over the past two and a half weeks. But there's something about one of the rocks. Maybe it's the boulder's chocolate brown color, or the fact that I hadn't noticed this particular rock before, among several familiar ones.

I shift my gaze to another part of the valley, then pull it back across the creek, and darned if that chocolate boulder hasn't moved. A couple of minutes earlier, it lay in an open meadow. Now it's beside a willow thicket, near the bottom of a gully. I hurry back to the tent and grab my binoculars.

By now the boulder has moved again, so I'm all but certain what I'm going to see. The glasses confirm it: a chocolate grizzly is grazing on tundra plants directly across Giant Creek, maybe an eighth of a mile away. Well, this is what I've wanted. But now, miles from the nearest human on a dark and dreary day, I'm not so sure.

The bear's shape suggests an adult: heavily muscled body, large belly, massive head. From here the grizzly easily looks large enough to be a male.

The bear works his way uphill, out of the thickets and back into the tundra meadow. I find myself encouraging this move: *Good bear; stay away from my camp.*

The wind and rain don't seem to bother him. In fact he moseys along, munching as he goes, as if he has no cares at all — which at this moment, and likely most moments in his adult life, is probably true.

I wonder what he's eating, since berries are scarce here. From what I can tell, he's consuming both greens and wildflowers. When the bear departs, I'd like to go inspect where he's grazing, but high, rain-swollen creek waters may prevent that.

Scribbling in my journal, I lose track of the grizzly. Ten or fifteen seconds pass before I relocate him, though he's not hiding. That's how well he blends in. Now he's lying among some cobble-sized rubble. An after-lunch nap? Out in the open, he's facing directly into the wind-driven rain.

The grizzly hasn't looked my way once that I can tell. I wonder if he's noticed me, or my purple-and-green tent, or my flapping beige tarp. From my own hikes, I know the tarp stands out, even from a distance. And now it's flailing in the wind. If the bear has noticed, what does he think or sense? What do his instincts tell him? Has he encountered humans before?

For the moment, at least, I'm much more interested in the bear than he is in me. That's good. That's what I want. Though I'm pleased to be sharing the valley with this grizzly, I wish to do so at a distance. Even having one this close to camp makes me a little nervous. One thing seems certain: the grizzly's experience of the valley is not enhanced by a human presence. Many residents of the nearest village would likely shoot the bear if they found him this close to camp, or even crossed paths with him on trips through the mountains.

The bear shows no anxiety about my presence.

It's *cold* out here. My fingers and toes are tingling. My small field thermometer reads in the upper forties, but with the gusting winds, the chill is probably closer to freezing: classic hypothermia weather. So here I am, behind a shelter and cloaked in wind shell, fleece jacket, Capilene shirt, nylon pants, wool cap with rain hood pulled over it, and fleece gloves — and starting to shiver. Meanwhile, the bear, lying on an exposed bench and facing into the wind, shows no discomfort at all. If he's not chilled at all on a day like this, does he overheat on the few bright, sunny, warm days of summer?

The bear naps more than an hour before rising. Immediately he resumes feeding. He does little else for the next two hours except chew mouthfuls of tundra greens. He barely lifts his head except to see where the next patch of food might be and seldom has to take more than a few steps to get it. He continues to ascend a rocky rivulet surrounded by low-lying but lush green plants, until reaching a rubbly pile of lichen-bearded boulders. He then tops a knoll that apparently has little to temp him and ambles to another lush swale, where he resumes his feasting. It would appear this grizzly has entered the late-summer phase that biologists sometimes call hyperphagia: almost around-the-clock gorging, in preparation for winter's fasting. At one point the bear finds a spot so luscious, all he can do is sprawl in the midst of it.

The more I watch this grizzly, the more I'm confident he's a heavily muscled, mature male, perhaps in his prime. He has little to fear, except a larger bear — which seems unlikely here — or people bearing guns. That's one good thing about this weather: no hunters are likely to come this way today.

Evening: I've finished dinner except for a few bites of dark chocolate and sips of coffee. My stinging-cold toes tell me not to stay out much longer. It must be five hours at least since I first spotted the boulder that turned into a bear. Over that time it's become clear that much of what he's eating is Richardson's saxifrage, also commonly known as "bear flower" — for good reason. The grizzly seems to be consuming it all: flowers, stems, and leaves. An eating machine. And havin' a ball. Again he sprawls out, head swiveling back and forth, mowing down those plants. It must be the closest thing to bear heaven, if you don't have access to berries or salmon.

Once, twice, the bear's attention is diverted. He stares down into the creek a few moments, then resumes eating. I wonder if backpackers might have come over from a neighboring valley. Or, even less probable, if Nunamiut hunters have come up here, unnoticed by me in the howling and drum-beating cacophony of wind and rain. "Don't you dare shoot that bear," I warn the imagined hunters.

I put away my food and cook gear, take some stuff back to the ten, then return for a final look. The grizzly is now on the move. Heading slowly toward the creek bottom, he occasionally stops for snacks along the way. I take down the tarp and retreat to the tent with my binoculars and sitting pad, setting up on the downwind side.

Just once that I can tell, the grizzly looks toward my tent. Then, still headed downstream, away from my camp, he disappears into thick brush that lines the creek. There's no sign of him for several minutes. Finally he reappears, going up a gully that leads to a side valley. Moving beyond shoulder-high willows, he heads onto open tundra, walking slowly but steadily with a gait that's part swagger, part waddle. He comes to a rise and tops it, body disappearing bit by bit till he's gone.

I've spent six hours, maybe more, in the company of a bear. It's one thing to do that at McNeil, with its guides and structured system. It's quite another while alone, deep in the wilderness.

I feel a vague emptiness as the bear disappears, a sadness to go with my delight at this gift, this extended peek into a grizzly's solitary life.

You could say nothing special happened today. The bear didn't do anything unexpected or threatening. My encounter with this grizzly won't grip friends and new acquaintances the way my Shuyak bear-attack story does. And I had no epiphany. Yet it's been everything I could have asked for and more.

There's little that could have kept me outdoors in such bleak weather, yet while I was engaged with the grizzly I largely forgot my own discomfort and moved more deeply into the raw power of this Arctic alpine valley. On Shuyak I was an unschooled intruder. Here I've slipped into the bear's world without threatening him or otherwise affecting his behavior, that I can tell. Though each on his own path, we've shared something here — a valley, a storm, an awareness of place.

I crawl into the wind-whipped tent, curl up inside my sleeping bag, and drift into a deep sleep. Body calm, spirits lifted, I move though these mountains with bear.

MICHAEL SHNAYERSON

The Rape of Appalachia

FROM *Vanity Fair*

THE IMAGES ARE still fresh: candlelit faces in the West Virginia night, family snapshots of coal miners trapped below, hopes of a rescue raised and cruelly dashed, the gentle letters written by dying men. The explosion at the Sago mine, in Upshur County, on January 2, 2006, was a national drama because thirteen men were buried alive while their families prayed above. But no less haunting was the glimpse it gave of an industry with just as much disregard for its workers as in the days of Dickens. Sago executives sat on the news of the miners' plight for a crucial sixty-nine minutes; the first rescue crews failed to arrive until four hours after the explosion. Sago's billionaire owner, Wilbur Ross, a staple of Manhattan society with his latest wife, Hilary Geary, touted his mines' safety record; the truth was that federal inspectors had issued Sago 208 citations in 2005, up from 68 in 2004, while state inspectors had issued it 144 citations, up from 74 in 2004.

Why had Sago been allowed to operate after all these infractions? Because in West Virginia, coal is king. Inspectors can write all the citations they please. Coal-industry lawyers will just pay them as a cost of doing business. Governors and senators can call for reform, but King Coal underwrites their campaigns. When Sago was followed, in less than a month, by four more mining deaths in the state, West Virginia governor Joe Manchin III stood up to demand, dramatically, that every mine in the state be shut down until deemed safe. How long would that take? A bantam figure beside Manchin named Bill Raney, president of the West Virginia Coal Association — the voice of the industry — supplied the answer. "Two hours, we think," he said. "Maybe four." One might have thought he was joking, but no.

With underground mining, the industry's arrogance is usually hidden, like the miners below and the coal they're prying loose from the earth. Not until a tragedy does its flouting of laws and regulations come to the surface. But every day in West Virginia, that arrogance is on display aboveground, in the grimly efficient, devastating practice known as mountaintop removal. Hardly any miners die as a result of it, though one of the four deaths after Sago did occur, freakishly enough, when a bulldozer operator on a mountaintop site hit a gas line that ignited and then set him afire. With mountaintop removal, it's the landscape that suffers: mile after mile of forest-covered range, great swaths of Appalachia, in some places as far as the eye can see, are being blasted and obliterated in one of the greatest acts of physical destruction this country has ever wreaked upon itself.

You can see the results all too clearly from a plane at 35,000 feet. You can see them in satellite pictures too. But you won't see a tree out of place as you drive south from the gold-domed capital city of Charleston into the low-lying mountains they call, incongruously, the coalfields. The coal companies may be brutish, but they aren't stupid. They site their operations well away from the interstates, even a ridge or two away from the county roads. The unsuspecting traveler rolls down I-77 admiring the forested mountains, little imagining that the range he sees is as much an artifice as a Hollywood backdrop.

The first clue that something is amiss comes with the turnoff from I-77 that leads to two-lane Route 3, in Racine. This is the road that snakes down the Coal River Valley, into the heart of the coalfields. The mountains, still intact, move closer in now, the valley so narrow and deep it forms a separate world. Few signs of the region's famous poverty are apparent. The houses, though humble, are well kept; the locals stopping for gas drive late-model pickups. Suddenly, a 60-ton coal truck hurtles down the curving road, its full load swaying precipitously over the double yellow line, its driver yanking the Jake Brake if the car in front doesn't move fast enough. A second coal truck roars into view from the other direction, and the car in between is rocked by its tailwind as the semi goes by: a little reminder of where power lies in the Coal River Valley.

The towns are small and neat, almost storybook, but something is clearly wrong with Whitesville, once the beating heart of the valley. It looks desolate, its storefronts abandoned, its streets and sidewalks still. Hardly a car is parked here, not a soul to be seen. Up and down the length of its Main Street, only two businesses appear to be thriving. Both are florists. Poor though West Virginians are, they buy a lot of funeral flowers. Whitesville resembles a wartime town pillaged by an advancing army. In a way, that's what it is.

You have to get up to a ridgetop to see that army's path. The view from Larry Gibson's place will do just fine. Gibson, a leprechaun of a man with a surprisingly deep baritone voice, lives on the top of Kayford Mountain, just east of Whitesville. His ancestors moved to the valley in the late 1700s and acquired five hundred acres of the mountaintop in 1886 through a wedding dowry. Twenty years later, a land-company agent from out of state gulled an illiterate forebear into signing his X on a contract that transferred most of the land for "one dollar and considerations." Almost everyone in the Coal River Valley has a story like that. The Gibsons, unlike most families, managed to keep fifty acres at the top of their mountain. This is where Gibson lives still. He's the only holdout for miles around, his mountaintop a little green island surrounded to the horizon by brown, raw, devastated earth.

This is what lies behind the picturesque landscape of roadside hills in the Coal River Valley: mountains reduced to rubble. The topography you can see from Gibson's mountaintop compound is so much lower that it's hard to imagine the forested ridges that once rose here. It's like a manmade Grand Canyon, except that the Grand Canyon teems with life and this panorama has none — none except the men who work the distant dozers and huge-wheeled dump trucks, their motors a constant, hornet-like hum.

The coal companies have tried hard to buy Gibson out because, he says, Kayford Mountain has thirty-nine seams of coal, worth $450 million, directly under his property. Gibson has turned them down. He believes they want him gone too, because he still bears witness to what they're doing here. That's rare. The coal companies own or lease nearly all the land outside the valley's towns, and for the most part they can fence off their operations, keeping people a ridge or two away from their mountaintop sites. Gibson's property, with its more than a dozen shacks and family cemetery, is a vantage point for journalists who want to see what mountaintop mining is about.

The coal companies hate that, Gibson says, and they find ways to let him know it. They tell their miners that Gibson is out to take their jobs away. In response, Gibson claims, miners have shot up his place when he was there; his trailer has the bullet holes to show for it. They've torched one of his cottages. They've shot one of his dogs and tried to hang another. They've forced his pickup off the road, tipping it into a ditch, and paused long enough to laugh at him as he tried to get out. Gibson keeps a growing list of all the acts of violence and vandalism committed against him and his property. Currently it totals 119. The stress of these threats, and of making his mountain a cause, led his wife to leave him not long ago. "If I stopped fighting for the land maybe we'd have a chance," Gibson says. "But this is my heritage. How can I walk away from that?"

Kayford is astonishing. But it's just one of nearly a dozen mountaintop-mining sites that ring the Coal River Valley. It's one of scores of sites in central Appalachian coal country. The U.S. Environmental Protection Agency, even while sanctioning the practice, concluded in 2003 that 400,000 acres — all rich and diverse temperate forest — had been destroyed between 1985 and 2001 as a result of mountaintop mining in Appalachia. That figure is probably 100,000 acres out of date by now. In those same sixteen years, the EPA estimated, more than 1,200 miles of valley streams had been impacted by mountaintop-mining waste — of those, more than 700 miles had been buried entirely. That figure is old now too.

This would never happen in rural Connecticut or anyplace where such destruction would stir universal outcry and people with money and power would stop it. But Appalachia is a land unto itself, cut off by its mountains from the East and Midwest, its people for the most part too poor and too cowed after a century of brutal treatment by King Coal to think they can stop their world from being blasted away. They're probably right: the EPA sees no reason to think that in the next ten years its statistics of damage by mountaintop mining won't double.

The story of mountaintop mining — why it happens, and what

its consequences are — is still new to most Americans. They have no idea that their country's physical legacy — the purple mountain majesties that are America — is being destroyed at the rate of several ridgetops a week, by 3 million pounds of explosives every day. They remain oblivious to the fact that along with the mountains, a mountain culture is being lost: a culture of families who, like Larry Gibson's, go back six, eight, or more generations — a community of deeper historical roots than almost any other in America today.

The reasons so much mountaintop mining is being done now are simply enumerated: money and politics. For decades, coal was the fuel of last choice, visibly dirty and messy, its black smoke an urban blight, its sulfur compounds and nitrous oxide known, even then, to cause acid rain and strongly suspected to be greenhouse gases partly responsible for global warming. Homes and buildings once heated by coal were converted to gas and oil, and scientists predicted that coal would soon be a fuel of the past. But half of the country's electric plants were still powered by coal, and power companies balked at the cost of abandoning it. Now that the cost of oil is so high and Middle East politics so fraught, coal is back. In fact, it is the centerpiece of President Bush's energy plan, because America is said to have 250 years of minable coal reserves, much of it in these Appalachian coalfields. More than one hundred vast new coal-fired power plants are being built across the country, and hundreds more in China. The price per ton has soared on the global market, from about $20 to more than $50. In the coalfields, these are boom times, and the best, most efficient way in Appalachia to satisfy the insatiable demand is to blow off the mountaintops to get to the seams that lie like layers of icing below.

It would be hard to imagine a more ill-advised course of action than ruining large swaths of land to get coal and then poisoning the atmosphere with the gases from burning it. Yet the incremental damage from mining and burning a billion tons of coal a year is hard for most Americans to see, so they don't worry about it. The 18,700 people in the U.S. who die each year of coal-dust-related respiratory disease do so singly, and invisibly. The downwind emissions from coal-fired plants dim views and kill aquatic life, but subtly, over time. The mercury produced by burning coal settles in waterways, to be consumed by fish and work its way up the food chain, but this too is an invisible process, and so the danger, particularly to pregnant women, is ignored.

Until recently, mountaintop mining went unnoticed as well, even by many Appalachian residents, and certainly by West Virginians north of the coalfields. That's changed. So voracious are the coal companies now, seizing their chance with an administration that does all it can to encourage them, that the mountaintop sites have broadened dramatically, and even local families whose fathers and grandfathers proudly mined underground coal have begun speaking out.

The Coal River Valley is pretty much the geographical dead center of this ruination in Appalachia. And here, as it happens, an archetypal drama is playing out that's both local and global. It pits Don Blankenship, West Virginia's most notorious and unapologetic mountaintop-mining coal baron — arguably its richest and most powerful figure — against a slew of critics. Among those trying to rein him in are a young environmental lawyer named Joe Lovett, a coal miner's daughter named Judy Bonds, and Governor Manchin.

Until last year, Blankenship was merely a man who, according to Cecil E. Roberts, president of the United Mine Workers of America (UMWA), had caused more misery to more people in Appalachia than anyone else. He had grown up poor in the railroad depot town of Delorme, West Virginia, the son of a single mother who supported four children by working long hours in a gas station grocery mart. The coal-filled hills around Delorme still seemed to echo with the early battles of miners against bosses and their hired hooligans; Matewan, scene of the epic 1921 showdown and subject of the eponymous 1987 John Sayles movie, lay just up the road. But as a young Massey Energy manager when the UMWA struck in 1985, Blankenship had sided with the company and come away from the eighteen-month standoff profoundly affected. Later, as chairman and CEO of Massey Energy, he made a practice of union busting as he acquired rival mines throughout the 1990s. Once, every mine in the Coal River Valley was a union operation. Now, thanks to Blankenship, hardly a union mine remains. All this growth has made Massey the fourth largest coal company in the country, with more mountaintop mines than any other — more than two dozen, many of which encompass thousands of acres. Nearly all the sites in the Coal River Valley and its environs are Massey mines.

Other coal companies — Peabody for one, Arch for another, Consolidated for a third — do mountaintop-removal mining too. But Massey does it with a vigor — displaced residents might say ruthlessness — that leaves its rivals shaking their heads. Typically, the first indication for residents of a hollow that their lives are about to be turned inside out by a new Massey mountaintop operation is a car driving up and down the winding hollow road, stopping so its driver can take notes about who owns which house. In an air-conditioned office, Massey lawyers will research who owns mineral rights, who owns just land, and who owns nothing at all. The ones who own mineral rights may have to be bought out. The rest can be ignored.

One day soon after that, a timber contractor clear-cuts a ridgetop where families in the hollow hunted and fished and hiked for generations. Then the blasting begins. Daily detonations of ANFO — a mix of ammonium nitrate and fuel oil, the same explosive that Timothy McVeigh used in Oklahoma City — cause reverberations that crack foundations and walls, destroy wells, and rack everyone's nerves. Coal trucks start barreling up and down the narrow hollow road. Coal dust from their open loads coats the houses and cars. At times, toxic chemicals spilled from the site above turn the hollow's streams black.

Somnolent as it often seems, the West Virginia Division of Environmental Protection still managed to issue 4,268 citations to Massey operations between January 1, 2000, and December 2, 2005. A violation is anything that looks amiss when an inspector comes by. By comparison, Arch earned 732 violations over that same period, with 355 for Peabody. Far more serious is a pattern of three violations for the same problem. For that, the WVDEP may summon the offender to a show-cause hearing — to explain why he shouldn't have his permit revoked. Over that six-year period, Massey had 117 show-causes. Arch had 20. Peabody had none. Jeff McCormick, of the WVDEP's Division of Mining and Reclamation, puts it bluntly: "Numbers don't lie."

For none of these activities or infractions does Massey reach out to the communities it's destroying with an explanatory or apologetic word. The way it communicates is by posting DO NOT ENTER signs on its mines' entry gates and by fencing off all the mountain land it owns or leases. Families in the Coal River Valley have lived with mining for nearly a century. But until Massey came, no one had kept them from the hills they called home. If they sue for some aspect of the damage done to their homes and land, they end up in court for years at great personal expense. If they try to move away, they find no one to buy their homes — except, sometimes, Massey, which might pay "fair market value" for houses in its direct path. "Fair market" means what the houses are worth now that a mountaintop-removal mine site is up the hollow.

Blankenship is all too well known wherever Massey mines, but until the autumn of 2004 he remained a local figure. That was when he spent $3.5 million to start a 527 political action committee — the newest loophole for unlimited campaign contributions, ironically produced by the McCain-Feingold Campaign Finance Reform Act of 2002 — aimed at toppling one of the five judges sitting on the West Virginia Supreme Court of Appeals. The money was a pittance for him: that fall alone, he cashed in $17.6 million in Massey stock options — notably more than the $13.9 million the company earned in profits that year. (Massey Energy is a publicly traded company, but Blankenship has a friendly board, which he dominates.) In West Virginia politics, though, a relatively small amount of money goes a long way. What happened next was an unabashed show of raw power — unprecedented in its openness — that served as an object lesson in why mountains keep getting blown up in the state despite overwhelming public disapproval of the practice.

The target of Blankenship's ire, Warren McGraw, was a proudly liberal justice who tended to side with coal workers in injury cases. A popular figure, he thought he'd skate to another twelve-year term, until billboards began appearing around the state with the cryptic message WHO IS BRENT BENJAMIN? McGraw's opponent was indeed unknown, but not after the signs and a statewide media-and-phone campaign, paid for by Blankenship. The coal baron's 527 PAC, ". . .And for the Sake of the Kids," beat into voters a message that McGraw let sex offenders free among their children. The campaign was based on the case of a juvenile delinquent named Tony Arbaugh, who after being sexually abused for years by his family had briefly abused a younger brother, done illegal drugs, and broken the terms of his parole. A lower court sentenced him to thirty-five years; McGraw, voting with the majority, gave Arbaugh one more chance and endorsed a lower-court minority suggestion that Arbaugh be given a janitorial job at a Catholic school. Later, McGraw said the specific suggestion to place Arbaugh at a school had been inserted into the opinion without his knowledge, and pointed out that Arbaugh had not, in any case, worked at the school after all. But the damage was done.

Blankenship's campaign made no mention of Massey Energy, or of the several appeals pending with the state Supreme Court in which Massey has a strong economic interest — including one where Massey is trying to reverse a $50 million judgment for having driven a small rival into bankruptcy. The campaign was a bullying push, and it worked. McGraw lost, Benjamin won, and Blankenship became a potent political force.

The elections also brought a new governor, and soon Blankenship took him on too. Joe Manchin III is a Democrat, but party labels mean little in a state ruled by the coal industry. It's more relevant to say he's a former coal operator, though perhaps a well-intentioned one. Just days after taking office, he proposed a bond issue to shore up pension funds for state workers. The plan relied on putting the money raised into the stock market, where, Manchin predicted, it would earn a consistent 4 percent. Blankenship saw that as a dangerous gamble. He may have been right. Having decided he liked politics, he spent hundreds of thousands more of his own money on a campaign against the bond proposal. When it came to a vote last June, the governor's signature issue went down in flames.

Was the governor, as a result, a bit more receptive than he might have been to angry cries from Coal River Valley residents about a Massey operation that might threaten the students and teachers of a school beside it? Or was he just doing his job to hear them out? Either way, his entry into this local issue has sparked a high-profile feud between the governor and his most powerful constituent. Ultimately, it's a standoff between the state and the coal industry — one the state may not win.

One day last summer as the feud was brewing, two men climbed to the top of a steep ridge to observe, surreptitiously, the action at a new Massey mountaintop site. Bo Webb, fifty-seven, is an ex-Marine who served in Vietnam. He has a tattoo that runs the length of one arm, depicting an Appalachian mountain range. Ed Wiley, forty-seven, was a football standout in high school and remains one of the valley's best turkey-hunters. Both are members of Coal River Mountain Watch, a local group that does what it can to stop the spread of mountaintop mining in the valley.

The side of the ridge the men walked up was thick with trees, but at the top they looked down at a big brown bowl of blasted land. Once, this was a valley with trees and houses and people, called Shumate's Branch. Now it's part of a 1,747-acre mountaintop mining site.

The men stayed nearly two hours, hidden amid the ridgetop trees, passing a pair of binoculars back and forth. They were watching tiny figures in the distance on an already flattened hilltop of the vast operation. Through the binoculars, Webb and Wiley could see them tamp ANFO into a grid of holes. Only a few figures were working — one reason mountaintop-removal mining is so efficient. Instead of the hundreds who labor in underground mines, these few set the blasts, then drive the bulldozers that push debris into the valley below. When they reach the seam, these same few men can operate heavy machines that scoop up 100 percent of the coal exposed. It takes just one man to operate the biggest machine of all: a giant steam shovel called a dragline, which can be up to 22 stories high and lift 100 tons of dirt per scoop. If one reason for Whitesville's demise is the environmental onslaught from mountaintop mining, another is the lack of jobs. The industry mines more coal than ever, but with a tiny fraction of the men it once employed.

At last the figures on the mine site got into tiny trucks and drove to a safe distance. There was an eerie silence, minutes long, and then the blast. It was strangely beautiful, a kind of performance art, sending sprays of dirt hundreds of feet into the air, where for a gravity-defying moment they lingered, before drifting back down to earth.

Coal from this site, Webb and Wiley knew, would be moved by conveyer belt over the next ridge, down to the Massey preparation plant behind the Marsh Fork Elementary School. It would be washed with chemicals at the plant to free it of rock and debris.

The cleaned coal would be stored in a silo 110 feet tall. Then it would be loaded onto open train cars bound for market.

All this would happen within 300 feet of the school.

Marsh Fork Elementary is a low-lying bunker of a building on Route 3 in Sundial, about 9 miles south of Whitesville. It's set back from the road by a wide front lawn and tucked into a curve of Marsh Fork Creek. Behind the creek, a mere stone's throw away, looms the fenced-off preparation plant of the Goals Coal Company, a Massey subsidiary.

You can see the school from Route 3, and the prep plant and silo behind it. What you can't see is the vast earthen dam, called an impoundment, built into the hillside, where the chemical slurry left over from washing the coal is stored. From a helicopter hovering above the valley, it looks like a huge, open-topped hornet's nest, filled to the brim with black, opaque liquid. Until a few years ago, most parents of Marsh Fork Elementary's 217 students didn't know it existed. They were surprised, the more so to hear that it holds 2.8 billion gallons of highly toxic liquid.

Impoundments are a big part of mountaintop mining, and they're troubling even to some supporters, such as Senator Robert C. Byrd (Democrat, West Virginia), who commissioned a study on their dangers in 2001. The slurry they hold is an environmental disaster in itself. But also, impoundments can and do rupture. Just over the Kentucky border, a Massey impoundment in Martin County ruptured in October 2000, sending 306 million gallons of sludge into the local waterways. The Environmental Protection Agency called it "the worst environmental disaster in the southeastern United States." In 1972, an impoundment break at Buffalo Creek, West Virginia, released 132 million gallons of sludge that drowned 125 people. In some cases, as in Martin County, impoundments are built over long-abandoned underground mines, and the growing mass of slurry eventually breaks through to the catacombs below and seeps out the mine mouth.

Officially, if a rupture occurred, an alarm would be sounded "personally or by bullhorn," and the children of Marsh Fork Elementary would be driven to schools down the valley. Unofficially, the children would probably be buried. Massey supporters in the area — which is to say teachers, parents, and other locals whose relatives work for the company — say this is a case of Chicken Little, since the impoundment was built in 1985 and no major rupture has occurred. (Small leaks of toxic liquid are considered routine.) Davitt McAteer begs to differ. The assistant secretary in the Department of Labor's Mine Safety and Health Administration under President Clinton, he headed up Senator Byrd's impoundment study and is now directing Governor Manchin's study of the Sago mine collapse. He points out that at the 110 impoundments in West Virginia built since 1972 there have been 33 spills or ruptures — more than half of them Massey operations — releasing, conservatively, 170 million gallons of sludge. "The fact that seventeen belong to one company tells you that this company is not dealing with the problems," McAteer says. "My bottom line is you shouldn't have a huge impoundment above a school, even if it's the best impoundment in the world."

In response to this and other points raised by *Vanity Fair,* Don Blankenship sent a brief letter. "The persons involved in producing, processing and shipping coal at Massey are lifetime residents of Appalachia who have no desire to pollute the environment or harm anyone," Blankenship declared. "The sources for your article are exemplified by Davitt McAteer. A man who did nothing for safety while he headed the Mine Safety and Health Administration. A plaintiff attorney . . . a man who says coal miners can be kept track of underground with twenty-dollar devices. Clearly, not a man to be believed."

For many Marsh Fork parents, a more immediate concern than the impoundment is the quality of air and water in and around the school and prep plant. Aided by Coal River Mountain Watch, several parents began comparing notes last year and claimed that their children share chronic symptoms. Herb Elkins says his seven-year-old granddaughter has a runny nose, a cough, and earaches all through the school year. Two weeks into summer vacation, she's fine. Bob Cole's son Davy has chronic stomachaches that get him sent home routinely, but his parents never get them. "They've had as few as four or five kids in a class, the rest out with diarrhea and vomiting," Cole says. "In the middle of the day they throw up." Carolyn Beckner's daughter Brittany has such bad stomachaches that she asks her mother nightly to pray for her stomach to feel better.

Kenny Pettry, a teacher at Marsh Fork, whose own ten-year-old son, Jacob, attends the school and has chronic upper-respiratory symptoms, believes the students grew sicker in 2003, when the silo was built behind the school. "The trains are loaded up right by the silo, so you're getting the clouds of coal, plus all the chemicals they're using to clean it with. What are they using to clean it with? Massey has never told us."

Just as worrisome as the air is the water. Several children have reported getting sick after drinking from the school water fountain. Pettry feels that blasting has damaged the school's septic system, from which rank odors frequently emanate. Don Price, Marsh Fork's current principal, however, says the air and water at the school are fine and that there have been no specific cases of children becoming ill from either.

As for Marsh Fork Creek, all it takes is one look to see it's fouled. Not long ago the stream was clear, local residents say. Now it's muddy brown. Down along its bed, a channel of darker brown flows like a slithering snake. Parents say children who swim in Marsh Fork or its tributaries come out sick now, with flushed faces, headaches, raspy coughs, sore throats, stomachaches, and diarrhea.

It was in the climate of these rumors and fears that parents reacted with such dismay to the news last year that Massey intended to build a second silo behind the school.

More and more coal was coming down from the mountaintop sites behind the ridge, Massey explained to the WVDEP. Another silo was needed to store it. The first silo was already less than 300 feet from the school; the new one would be just 260 feet away. The federal surface-mining act of 1977 prohibits coal companies from building within 300 feet of a school, but Massey had an answer for that: its prep plant on Marsh Fork Creek had been operating since 1975, making it exempt from the law. So confident was Massey that it went ahead and built the foundation for the 168-foot silo while the permit was still pending. Generally, what a coal company in West Virginia wants, it gets.

This time, though, a reporter from the *Charleston Gazette* upset Massey's plans. Ken Ward Jr. has covered mountaintop-removal mining for nearly a decade with a grim determination and thoroughness that make his stories the ongoing chronicle of this national tragedy. He might have moved on long ago to the *New York Times* or the *Washington Post,* but like Larry Gibson, he stays to bear witness. When Massey declared that its silos were exempt from the 1977 law, Ward started studying the site's permit applications over the years. What he found was extraordinary. Neither of the silo sites was entirely within the originally permitted area, but on subsequent Massey survey maps the property lines had *migrated.* They'd moved toward the Marsh Fork Elementary School, just far enough to accommodate the first, and now the second, silo. Ward's revelation startled the WVDEP, which had just granted a permit for the new silo on the basis of Massey's own maps. And it galvanized the governor. Now, when the agency looked again, it did something no one could remember it ever having done before: it rescinded the permit.

Blankenship was furious. The governor, he declared, was simply taking revenge for the coal baron's campaign against the bond issue. And so, by this logic, Blankenship sued the governor for violating his right to free speech. It was like two soldiers shooting each other at close range, Blankenship said. "He shut my silo down, and I shot him with this lawsuit." He waved off the matter of the migrating boundary line as a technicality "much like being a mile over the speed limit. They could apply this to every coal company in West Virginia and probably shut them all down," Blankenship said. "If all of us are going fifty-six miles per hour and I'm the only one getting a ticket for speeding, I have to be concerned about that."

In the Whitesville storefront office of Coal River Mountain Watch (CRMW), Judy Bonds heard about the WVDEP's reversal on the silo permit with mixed feelings. She felt it averted an immediate threat to the school, but parents were still concerned for their sick children, and the impoundment still loomed behind them. The governor had told her in his office that he would take action, but so far nothing had happened. Only Ken Ward's story had made the WVDEP rescind its permit. Even now, Massey might prevail. This was no time for celebration.

A daughter and granddaughter of coal miners, Bonds came to CRMW five years ago, when a Massey operation blasted the ridgetops above her hollow and everyone had to move. She was the last holdout, stubborn and angry, but she was finally forced to go. At CRMW she found her voice as a public speaker. Now she's a statewide symbol of local resistance to mountaintop mining and the North American winner of the 2003 Goldman Environmental Prize, a prestigious international award given annually to one preservationist per continent. At fifty-three, she's a single mother, thrice divorced, with curly dark hair and shoe-button eyes that blaze with indignation when she starts in on what King Coal has done to the valley. Aside from her family, CRMW is her life.

To date, Bonds and CRMW haven't actually stopped a mountaintop-mining operation. But they track every permit and force public hearings, and that's progress: when she started, no one even knew when a permit was granted, much less if residents had a right to question it. Bonds and her fellow activists — Bo Webb, Ed Wiley, and half a dozen others — prod the press, wring meetings with the governor, and stir neighbors into speaking out. As news broke of Massey's migrating property lines, Bonds helped push the state to inspect the Marsh Fork school. (After checking for carbon monoxide, carbon dioxide, and mold, inspectors concluded the school was safe.) Then she took another step: she had CRMW sign on to Joe Lovett's suit against the WVDEP.

Lovett is a lone gun on the legal front here, the one local lawyer who's fought mountaintop mining in any consistent fashion. Boyish and intense at forty-seven, with clean-cropped brown hair, wire-rimmed glasses, and the blush of youth in his cheeks, Lovett might pass for a student at a New England prep school. His father was a prominent lawyer in Charleston, but Joe seemed to need to rebel: he spent much of his twenties on an organic farm. Environmental law was a kind of rebellion too. When he graduated from law school, at thirty-six, Lovett found himself listening, on his first day of work at a Charleston nonprofit law firm, to the plight of a coalfields resident whose hollow was about to be destroyed by a mountaintop-mining operation. Lovett had never seen a coal mine. He'd never even driven south of Charleston into the coalfields. That seems a long time ago now.

Lovett works behind a glass-paneled door with lettering that reads, APPALACHIAN CENTER FOR THE ECONOMY AND THE ENVIRONMENT, in a two-room office in Lewisburg, a gentrified colonial town an hour's drive east and a world away from the coalfields. A nonprofit group of his creation, modestly funded by a few donors, the center allows Lovett to sit at a cheap computer, framed by posters of Louis Armstrong and Robert Frost, and bat out crisp legal broadsides. This one went to WVDEP secretary Stephanie Timmermeyer, on behalf of Coal River Mountain Watch. He demanded not only that the second silo be prohibited for good — no new permit application — but that the first one be torn down as well. Otherwise, he promised, he would sue.

Lovett's legal letters are written with cool authority, rarely revealing his rage against mountaintop-removal mining and, for that matter, the whole business of mining and burning coal. In the eight years he's been bringing coal cases, he's had remarkable success. The fact that his cases keep getting reversed on appeal is frustrating, though not surprising to someone as privately pessimistic as he. He knew from the start what he was up against, and yet he's kept going.

For Lovett, the silo case was irresistible, but untypical. He hardly ever takes on a particular case of alleged coal-company malfeasance. Instead, he looks for ways in which the whole industry breaks the law by mountaintop mining, aided and abetted by the state and federal agencies assigned to oversee it. So nettlesome is Lovett that the Bush administration, in trying to please the coal industry, has marshaled the full force of the federal government just to stop him.

At the heart of the administration's loyalty to coal is a simple political fact: West Virginia is a swing state with five electoral votes. In 2000, if it had gone to Al Gore, there might have been no aftermath in Florida, no endgame in the U.S. Supreme Court. Gore would have won the election. Coal delivered West Virginia to Bush by spending far more money than it had ever spent in a presidential race — $3.6 million, three times more than in 1996. Merely stopping Gore wasn't reward enough for the coal industry's investment. King Coal wanted payback — specifically, that Bush undo the damage caused by Joe Lovett.

In the Clinton years, Lovett had sued the U.S. Army Corps of Engineers for granting mountaintop-mining permits in a lax — and illegal — fashion. The Corps is best known for building bridges and dams, but when the Clean Water Act was passed in 1977, Congress gave the Corps a new job. It ordered the agency to pass judgment on any project that might damage U.S. waters. The Corps agreed, but only grudgingly. Ever since, it's done as little as possible to fulfill that responsibility in regard to mountaintop mining. In Charleston, a conservative federal judge, the late Charles H. Haden II, startled the Corps — and the mining industry — by saying as much, finding in Lovett's favor not just once but twice. But then Bush took office and proceeded to name an industry lobbyist to be the government's highest-ranking overseer of mining.

J. Steven Griles, who left as the Department of the Interior's deputy secretary in late 2004 to become a lobbyist again, may not have had a hand in all the rollbacks to rules that govern mountaintop mining, but he certainly emerged as a symbol of them — and of his industry's clout in squashing Lovett's court victories. Lovett and his legal partner, Jim Hecker of the Trial Lawyers for Public Justice, had stopped coal operators from dumping mountaintop waste into valley streams by claiming it was "fill" — a term in the Clean Water Act intended to mean sand and gravel that filled in a wetland for building, not mountaintop waste. Soon after he was confirmed as the number-two man at Interior, Griles touched on the fill issue in a speech to industry executives, saying, "We will fix the federal rules very soon on water and spoil placement." In May 2002, the EPA and the U.S. Army Corps of Engineers rewrote the definition of *fill* to include virtually everything except household garbage.

Technically, at least, a buffer zone of 100 feet on either side of streams protected them from being buried. Within that buffer zone, no mining activity was allowed unless a coal company determined, among other things, that a stream's aquatic life would survive the activity. The rule was often ignored, but at least it was there. At the Office of Surface Mining (OSM) in Bush's first term, a formal proposal was made to change that rule. Now coal operators would need only to show that they'd taken measures "to the extent possible" to observe the buffer zone and protect streams. Griles says he "provided no policy direction, and never reviewed or commented on the proposed rule." But as Interior's number-two man, he helped oversee all the department's agencies, of which the OSM is one. Lovett and Hecker sought an injunction against the rule change as soon as it was proposed. The Bush administration fought them in court and won on appeal. The rule is still pending.

The Clinton administration had agreed to do the first serious study of the impacts of mountaintop removal — an Environmental Impact Statement, or EIS. This too was in response to one of Lovett and Hecker's legal victories. In an interoffice memo of October 2001, Griles emphasized that the EIS should focus on how to streamline the government's permitting process for the practice. When the EIS was done, scientific studies of the environmental impact were pushed down to footnotes or appendices; the thrust of the 5,000-page study was how to make mountaintop mining easier to do, in blithe disregard of the devastation documented in the EIS's own studies. Griles denies he had anything to do with finalizing the EIS, but the agency that ended up leading the effort was, again, the Office of Surface Mining, under his aegis.

In thwarting Lovett, the industry has gotten at least as much help from the higher courts as from the administration. Three times, after favorable rulings from West Virginia federal judges, Lovett has been batted down by the Fourth Circuit Court of Appeals, in Richmond, Virginia — the most conservative court in the land. West Virginia is one of five states whose federal appeals are heard by the Fourth Circuit (the others are Virginia, Maryland, North Carolina, and South Carolina), so any case that Lovett brings in his home state is destined to travel, on appeal, to the block-sized granite-and-limestone building where, by southern tradition, justices conclude their sharp and often acerbic questioning by stepping down to shake hands with the lawyers from both sides.

Last September, when Lovett walked into the appeals court building in Richmond for the third time, he almost felt optimistic. His case, once again against the U.S. Army Corps of Engineers, seemed his most solid yet. Lovett felt all the justices had to do was read the clear, simple language of the Clean Water Act to see how the Corps was violating its duty — and the law. Inside the building, Lovett's heart sank. There on a bulletin board were the names of the three judges randomly chosen, from the circuit's field of fourteen, to hear the case. How could this be? Two of the court's most conservative judges, J. Michael Luttig and Paul Niemeyer, were on the panel — again. Both had been on Lovett's first two appeals too! And both had ruled against him each time. What were the odds against the same two judges being randomly chosen three cases in a row? Lovett and Hecker did their best, but in the hallway outside afterward,

Lovett was grim. "They're going to find against us," he said. "I don't know how, but they'll do it." He was right: just before Thanksgiving, the Fourth Circuit ruled that the Army Corps could handle its permitting any way it liked.

Almost as a lark, Lovett requested an *en banc* hearing. On rare occasions, a circuit's full field of judges may decide to reconsider a ruling made by one of its three-judge panels. In mid-February, in a five-to-three vote, the Fourth Circuit decided not to reconsider. But in issuing that decision the court acknowledged that five of its fourteen judges had recused themselves from hearing the appeal last fall. The reason: all five had financial interests in mining. That, explained a court clerk, was why two of the other judges had ended up on all three of Lovett's appeals.

So Don Blankenship and the mining industry, activists now feel, have little to fear from any case brought against them in a West Virginia federal court. Even if they lose, the case will go on appeal to the Fourth Circuit, where five of the circuit's fourteen judges will recuse themselves and two of the remaining ones have declared that mountaintop mining can be permitted as the overseeing agencies see fit. Now that Blankenship's candidate for the West Virginia Supreme Court is presiding, the coal baron can hope for more pro-industry rulings there too. Just to be sure, he's declared he'll finance another 527 to go after the five-member court's other reliably liberal justice, Larry Starcher, if Starcher runs for reelection in 2008.

Blankenship even seems to have neutered the governor. Last fall his lawyers demanded copies of any and all communications between Governor Manchin and the WVDEP in regard to the silo squabble over Goals's prep plant behind the Marsh Fork Elementary School. Blankenship argued that Manchin urged the WVDEP to rescind its permit as retaliation for Blankenship's campaign to kill the governor's pension plan. In mid-March, a county judge granted Blankenship's request. "The choice is clear," the judge ruled, "and it is in favor of open government."

At the same time, Massey's lawyers went before the state Surface Mine Board, a five-member panel that tries to settle mining disputes out of court, to argue that the silo permit was wrongly rescinded. Lovett went up against them and got a Massey engineer to admit he had in fact altered the boundary line on Massey's map of the plant beside Marsh Fork Elementary School. But he'd done it, he said, because of changes in the terrain, "not . . . to allow me to put silos in where people think they ought not to be." The Surface Mine Board was unimpressed. It ruled that the WVDEP had every right to rescind the permit. Now Blankenship will take that ruling to court. If the court-ordered disclosure of Manchin's communications with the WVDEP shows anything embarrassing, he just may win.

Lovett keeps filing lawsuits: he's got a new one against the Army Corps, in a Kentucky court this time, so that when the case is heard on appeal, it will be taken up not by the Fourth Circuit but by the Sixth, which oversees Kentucky, Michigan, Ohio, and Tennessee. In Whitesville, Judy Bonds and the others at CRMW continue to speak out, angrier than ever. But it's hard to keep faith in the valley, where even when you win, you lose.

Just 3 miles north of Whitesville lies the once pristine town of Sylvester, population 195. A few years back, Massey Energy built the Elk Run prep plant at the edge of town, to process coal belted in from various mountaintop sites in the area. The plant and the trucks that began driving by in an endless cavalcade caused a black cape of coal dust to spread across the town, so that Sylvester's residents had to clean their windows and porches and cars every day, and keep the windows shut. Their town was ruined, and they had nowhere to go: no one would buy their houses now. In desperation, Pauline Canterberry and Mary Miller, two housewives in their seventies, led a town crusade to take Massey Energy to court.

The "Dustbusters," as Canterberry and Miller were soon known, saved coal-smeared paper towels in dated plastic bags and photographed the bags to prove each date. Represented by Charleston lawyer Brian Glasser, they won $1 million in damages for the town. Massey was forced to curtail the truck traffic; to put a dome, like an indoor-tennis-court bubble, over the dust-spewing part of its prep plant; and to install monitors around town and keep coal-dust emissions below a certain level.

Two years ago, Massey installed the monitors, but Sylvester is far from dust-free. Every day, as they did before the suit, the Dustbusters wipe down their windows and porches and cars. They talk all the time about going back to court.

That's the way stories end in the Coal River Valley: with a whimper, followed by a bang of blasting. Perhaps Don Blankenship won't get his second silo behind the Marsh Fork Elementary School, but most locals think he will. Then perhaps the increased activity of the site will lead the county to close the school, and the 217 children who attend it will have to be bused to a school a half-hour or more from their homes, on winding Route 3. That's what has happened already to one after another of the Coal River Valley public schools. And then Massey's subsidiaries will be able to build as many silos as they want.

They'll likely want quite a few, because all around Whitesville the Massey mountaintop sites are metastasizing: more blasting, more coal, more trucks and coal dust. Even now, a dragline is being constructed — at a cost of millions — to broaden the sites that Larry Gibson looks out on from Kayford Mountain. Locals believe the long-range plan is to depopulate the valley: to make life so unlivable in and around Whitesville that everyone leaves, the lucky ones with a Massey buyout check in hand, the rest without. Then Massey will be able to blast without worrying about activists such as Judy Bonds, or lawyers such as Joe Lovett, or even governors, because no one will be there to witness or care, and hardly anyone outside the coalfields will know that a sizable chunk of the American landscape is gone.

MEREDITH F. SMALL

First Soldier of the Gene Wars

FROM *Archaeology*

LUCA CAVALLI-SFORZA has recently decided to spend half the year in his urban house in Milan and the other half on the sprawling campus of California's Stanford University. The jet-setting lifestyle is nothing new to the eighty-three-year-old anthropological geneticist; he used to alternate months in each place, flying back and forth between continents as if it were a drive across town.

Cavalli-Sforza's migrations also exemplify the kind of movement that has been the main focus of his research for more than five decades. He has used the genetics of modern people to trace the path of our grand diaspora out of Africa about 200,000 years ago into every corner of the globe. On this journey into wildly different environments, our kind didn't speciate, but we did end up forming groups or populations that are distinct in skin color, body size, blood groups, and tendency toward certain diseases. And in those differences, Cavalli-Sforza reads our history.

His interest in human variation has also gotten him into deep trouble. He's a notorious figure among anthropologists because he has courted public attention and controversy. He has consistently maintained that there are no such things as biological races but at the same time has used racial categories to parse human types over evolutionary history. His career has been devoted to understanding the historical movements of indigenous groups, but some of the very people he studies have labeled him a racist and a "biopirate."

But all the controversy and name-calling in the world doesn't seem to have dampened his enthusiasm for the story of human history one little bit.

"Usually I ride my bike to work because I live on campus," Cavalli-Sforza explains to me with an energetic smile after he pulls up to his Stanford University office in a lipstick-red Mercedes-Benz sedan. We walk briskly into the medical school building and past his genetics laboratory, where Peter Underhill and Alice Lin are conducting ongoing research on the human Y chromosome for population studies, and into a small office crammed with piles of books and journals from more than fifty years of work. Cavalli-Sforza is in the middle of an analysis of the correlation of the genetic distance between populations and their geographic distance, and the desk is piled high with papers — clearly, the word *emeritus* has no meaning here.

"We are a successful species," he begins as we settle in to chat about what makes humans special. "After we are born — in the Western world at least — we have a ninety-nine percent chance of reaching maturity." And, Cavalli-Sforza points out, we are a species always on the go. Very quickly, he rolls out the history of French Canadians, the mix and match of people who became the Boers in South Africa, and the migration of Ashkenazi Jews. Each story is interwoven with the comings and goings, marriages and matings, of all sorts of people.

Cavalli-Sforza's history is almost as complicated. He trained as a medical doctor at the University of Pavia in Italy, but while he found medicine intellectually stimulating, the inability of science at the time to cure the majority of difficult cases was frustrating, and he eventually turned to genetics. Early in his career, he was interested in using genes to answer archaeological questions. First he became fascinated by the megaliths of Italy and began to think about the invention and spread of cultural practices, especially settled agriculture. Did people pass farming traditions from one neighboring community to another like gossip, or did people move, put down new roots, and bring this practice with them?

"Or maybe they stole it," he says, laughing. "There is a Basque legend that the Basques weren't initially farmers. But when farmers moved close to the Basques, they organized a feast and invited the farmers to dance. The Basques wore long socks, so the legend goes, and as they danced seeds from the farmers fell on the floor, stuck to the Basques' socks, and that's how the Basques brought the seeds home and began farming themselves."

In the 1970s, Cavalli-Sforza and archaeologist Albert Ammerman of Colgate University hypothesized that agriculture spread not by cultural diffusion but by farmers moving from one place to another, taking their practices along with their luggage. Plotting sites, they started in the Middle East as Ground Zero for the invention of agriculture and then mapped in all directions. In their view, farming communities grew, and when local agriculture could no longer support a burgeoning population, families, or grown children of farmers, moved on. In other words, Cavalli-Sforza and Ammerman were mating archaeology with the methods of population genetics for the very first time. "Initially, many people in archaeology thought this was an impossible combination," says Ammerman. "Even in the early 1970s, when there was considerable scope for big thinking in prehistoric archaeology, what we were trying to do was seen as a radical move."

Early on, Cavalli-Sforza also realized that although artifacts and excavations of living spaces are the primary clues to understanding past human behavior, the genes of modern people would also contain the path of history.

Each of us has genetic markers that we inherited from ancestors. These markers have been mixed through marriage, carried across mountains, sailed to foreign ports, and dragged over continents. And here they are today, genetic artifacts of individual history, In other words, our genes are biological potsherds left in our chromosomes, and they too can be unearthed, analyzed, and compared with similar material from other populations.

Cavalli-Sforza felt that he could bolster the farming hypothesis by looking at genes. At that point, geneticists did not have DNA itself in their toolbox, but for decades people around the world had been bled and typed for the ABO blood group, Rh factor, the human leukocyte antigen genetic markers. Cavalli-Sforza, Paolo Menozzi of Parma University, and Alberto Piazza of Turin University used these in modern populations to map the historical spread of populations from the Middle East into Europe, just as Cavalli-Sforza and Ammerman had used archaeological evidence. In the end, the geneticists came up with the same answer for the spread of agriculture — farmers and not the practice of farming moved about.

Those results also encouraged the three geneticists to begin work on their 1994 monumental tome, *The History and Geography of Human Genes,* which includes every possible reference to every known genetic variation in every group on earth. More remarkable, the researchers tried to make sense of how and why genes appeared in certain places; they combined archaeology, linguistics, and known genes to map who went where.

That step — the melding of modern genes, archaeology, and linguistics with the statistical methods of population genetics — established Cavalli-Sforza as an anthropologist in the grand tradition, a researcher who reached out beyond his own discipline to ask and answer questions about the "big picture" of human evolution. Why have humans been so much on the move? What was the path of our movement? And even though humanity did not speciate during our diaspora across the globe, what particular forces of evolution shaped how we look and behave?

To Cavalli-Sforza's surprise, asking and answering those questions eventually landed him in a quagmire of controversy that dogs him to this day.

The 1980s and 1990s were heady days in genetics, with the invention of all sorts of ways to duplicate and slice DNA in the laboratory. Although the revolution in molecular biology was aimed primarily at medicine and human disease, Cavalli-Sforza was quick to realize the potential for studying human evolution. He knew it would be possible to collect, replicate, and store strands of DNA full of endless information about a person's heritage. Instead of many blood proteins to compare, there would be thousand and thousands of genes to sort through and compare. In other words, population genetics was coming out of the Dark Ages and into the Renaissance, and Cavalli-Sforza was right there with the same questions that had been preoccupying him for decades.

In 1991, on the heels of the Human Genome Project, which aimed to map every human gene, Cavalli-Sforza conceived the equally ambitious Human Genome Diversity Project (HGDP). Instead of focusing on which genes make a human, the project aimed to figure out what makes humanity in all its variation and similarity. In a letter to the journal *Genomics,* Cavalli-Sforza and geneticists Allan Wilson, Charles Cantor, Robert Cook-Deegan, and Mary-Claire King proposed the collection of DNA from all corners of the globe, with a focus on those isolated groups on the edge of extinction.

"Population growth, famine, war, and improvements in transportation and communication are encroaching on once stable populations," they wrote. "It would be tragically ironic, if during the same decade that biological tools for understanding our species were created, major opportunities for applying them were squandered."

The researchers saw the HGDP as a scientific savior, a way to preserve samples of our past before these populations disappeared. And they saw nothing wrong with going out and getting what was needed. After all, researchers interested in human history have always collected data under the gun — ethnographers have rushed to record the ways of hunters and gatherers before they settled down, archaeologists have moved quickly to excavate and secure sites before the bulldozers arrive, and linguist run for their tape recorders to save the last remnants of a native language or dialect.

The HGDP initially focused on 700 of the estimated 5,000 populations around the world. Some, such as Mbuti pygmies, were at risk of disappearing and combining with other cultures, while others, such as the Japanese, were robust populations. Hoping for $25 million from federal agencies, they expected to send teams all over the world to those isolated groups to bring back blood samples to various labs where the cells would be cultivated (that is, reproduced) and DNA extracted from white blood cells. Blood samples would also be stored so that future researchers might forever dip into the freezer and thaw out a never-ending line of DNA from people who might be long gone.

From Cavalli-Sforza's point of view, this ambitious project was the mother lode. With DNA from all over the world, population geneticists could map the movement of ancient peoples from one place to another, get a handle on human variation, and figure out which forces of evolution were most important in forming our species. A human gene bank would also help medical researchers understand why certain diseases find purchase in some groups but not in others. The samples would be available to any researcher with an interesting hypothesis and responsible credentials.

From 1991 to 1994, there were four symposia supported by the

National Institutes of Health (NIH) and National Science Foundation (NSF) in which methods were batted about, but it soon became clear that the researchers had made gross miscalculations with the methodology, goals, and ethics of the HGDP. By 1994 there was so much furor that the agencies that had supported the initial symposia (NIH and NSF) asked the National Research Council (NRC) to review the project before anyone proceeded. In its 1997 report, an NRC committee made recommendations to address the ethical issues. They also commented that those involved didn't seem to agree on the goal or the methodology of the HGDP, and the whole endeavor was problematic. They chose not to endorse the project.

But the biggest miscalculation concerned the feelings of the very people the HGDP intended to study. Project researchers assumed that indigenous people would easily offer up their blood just at the very point when those groups were becoming globally connected, politicized, defiant, and resistant to anything that smacked of Western science.

Organized indigenous groups saw their people once again being exploited by the West and quickly rallied their forces. The U.S.-based Rural Advancement Foundation (RAFI) — now called the Action Group on Erosion, Technology, and Conservation — accused the HGDP scientists of "biocolonialism" and "biopiracy." Other international grassroots organizations dedicated to protecting the rights of indigenous people followed suit.

"The Vampire Project [the HGDP] is legalized theft," claimed the Central Australian Aboriginal Congress in 1993. "The Vampire scientists are planning to take and to own what belongs to indigenous people . . . We must make sure that our people are not exploited once more by corporations, governments, and their scientists."

The National Congress of American Indians wrote, "NCAI does hereby condemn the HGDP and call upon all parties and agencies involved in it and related activities to cease immediately."

"The collection of genetic samples from indigenous peoples such as the Human Genome Diversity Project is unethical and immoral, and must be brought to an immediate halt," echoed the Maori Congress of Indigenous People from New Zealand. Following up, the Indigenous Peoples Council on Biocolonialism drafted ten resolutions condemning the HGDP and calling on funding agencies to pull out of the project.

The collective protest of these groups was understandable. For generations their people had been poked, prodded, bled, measured, and photographed, all in the service of Western science and with no recompense.

Cavalli-Sforza, however, believes the controversy was stoked by various organizations whose livelihood depended on "supporting indigenous peoples by representing them in mostly correct fights against multinational corporations that are exploiting them." By generating controversy, the organizations increased their income from sympathetic donors, the scientist believes. "It was impossible to convince them that we are honest people, and there was nothing that indigenous peoples would lose, maybe even gain something, by our research," he adds.

More surprising, many scientists also protested the HGDP soon after it was conceived. The backlash within the academy was led by molecular anthropologist Jonathan Marks, now at the University of North Carolina, Charlotte. Marks was vocal with anthropological colleagues — so vocal, he claims, that his opposition let to his removal from symposia on the HGDP — and in 1995 he first made those criticisms formal in the American Anthropological Association's *Anthropology Newsletter.* Marks pointed out that the HGDP was based on an old-fashioned idea about human groups, useless to medicine because there would be no record of medical histories, arrogant in its approach to indigenous people, an greatly confused about how to construct evolutionary trees from genes.

"Evolutionary trees are sensitive to phylogenetic relationships, patters of contact, demographic expansions and contractions, clustering algorithm used, and the individuals and populations chosen to stand for the larger groups," Marks says. "To pretend that they are just unproblematic representations of phylogeny is a mistake." Marks also saw the HGDP scientists as incredibly naive.

"The population geneticists had gone to bed with a nice idea, the amassing of a large repository of data on the population of the world, and had awakened on the cutting edge of contemporary anthropology and bioethics. They were entirely unprepared for it," wrote Marks in his book *What It Means to Be 98% Chimpanzee.* Marks also says that anthropological geneticists were also blinded by the possibility of easy funding, and that they had never been all that ethical. "The anthropological geneticists had been flying under the bioethical radar for a long time until the HGDP shined a bright light on their practices," adds Marks.

Cavalli-Sforza is bluntly dismissive of the criticism. "Jon Marks does not know the methods or the mathematics behind [the HGDP]. His own research on the primate family tree at the beginning of his career was a disaster."

Ken Weiss of Pennsylvania State University was brought into the HGDP as an anthropologist, but eventually left. He still supports the idea of gathering genetic data around the world: "I think that some kind of permanent, well-sampled, openly available source of DNA and genetic data was and is important." And Weiss points out that that goal has been accomplished even if the HGDP never really got off the ground. "There are now many sources of variable density data on human variation, including Hap-Map, spin-offs of the Human Genome Institute at NIH, and the Allele Frequency Data Base at Yale." Geneticist Kenneth Kidd of Yale University, who was involved in the planning stages of the project, also feels that Cavalli-Sforza can take some credit for those collections. "While I don't think it's quite cricket to call any of this the HGDP, the spirit and reason for the collection of such data has not changed. So Luca and a small band of his colleagues deserve high credit among anthropologists," he adds.

Over the past decade, Cavalli-Sforza has struggled to keep the idea afloat, and he defends the HGDP today. There are currently 1,064 cell lines in a laboratory at the Center for the Study of Human Polymorphism at the Foundation Jean Daussat in Paris, all collected by Cavalli-Sforza himself or by his friends for some other research project. According to a recent review article of the HGDP in the journal *Nature,* fifty-six laboratories have requested these materials since 2002, and there have been several publications on HGDP data that have underscored the force of geographic isolation and genetic drift (chance) in forming genetic differences among groups.

Cavalli-Sforza is also chairing the advisory board of the Genographic Project, which aims to do almost exactly what the HGDP proposed more than a decade ago. Headed by geneticist Spencer Wells, who once worked with him at Stanford, and supported by $40 million from the National Geographic Society and IBM, the Genographic Project is also poised to gather DNA from around the world. Of course, they've had time to learn from the mistakes of the HGDP. To smooth relations with the public, for example, Wells has invited anyone to contribute their DNA by purchasing a kit for $100 and sending in a swab of cheek cells. In an attempt to dispel the thought of "biopiracy," the Genographic Project is also sticking to collecting DNA and will not store cell lines that could be reproduced over time. Cavalli-Sforza says that the Genographic Project is almost the same as the HGDP, but Wells has the advantage of learning from others' mistakes.

But still, Wells is currently running into the same charges of "biocolonialism" as Cavalli-Sforza. The Indigenous Peoples Council on Biocolonialism says the Genographic Project is just a renewed version of the HGDP and its objections to both projects stand.

Cavalli-Sforza dismisses these protests against the HGDP and the Genographic Project as mere ignorance. "There are some people who hate biology," he says, waving a hand as if to keep the dissenters at bay. "Or they hate humanity."

Marks counters that Wells is blindly moving into the same deep waters as the HGDP. "The Genographic Project is an end run around the anthropological and bioethical issues that snagged the HGDP, and a black eye for National Geographic and IBM," says Marks.

Cavalli-Sforza also feels that the protesters misunderstand the intention of the projects, and he is adamant that no one is trying to steal, take, or victimize anyone. They only want to understand the genetics of human populations, and to him, such a goal is simply interesting science.

"There is a large section of people who are always afraid of science, and they are afraid of many things," he points out. "People are afraid of science because it is powerful, and it's unknown to them. People are afraid of the unknown."

A half-century of studying humans has left Cavalli-Sforza philosophical about the large and small movements of people across the globe. "People move because of economic reasons, to find food, or from fear of enemies. And sometimes women move because they need mates," he says. According to him, our ancestors moved, then they stayed in place for a while and let natural selection and chance decide which genes were passed down through generations and which ones would disappear. And in this evolutionary shuffle of human genes, Cavalli-Sforza also sees a natural, biological exuberance that has served humanity well.

"You never know what will happen, and so let us keep a lot of diversity among humans," he says, smiling again and gathering papers in anticipation of his next migration to Italy.

ROBERT H. SOCOLOW AND
STEPHEN W. PACALA

A Plan to Keep Carbon in Check

FROM *Scientific American*

RETREATING GLACIERS, stronger hurricanes, hotter summers, thinner polar bears: the ominous harbingers of global warming are driving companies and governments to work toward an unprecedented change in the historical pattern of fossil-fuel use. Faster and faster, year after year for two centuries, human beings have been transferring carbon to the atmosphere from below the surface of the Earth. Today the world's coal, oil, and natural gas industries dig up and pump out about 7 billion tons of carbon a year, and society burns nearly all of it, releasing carbon dioxide (CO_2). Ever more people are convinced that prudence dictates a reversal of the present course of rising CO_2 emissions.

The boundary separating the truly dangerous consequences of emissions from the merely unwise is probably located near (but below) a doubling of the concentration of CO_2 that was in the atmosphere in the eighteenth century, before the Industrial Revolution began. Every increase in concentration carries new risks, but avoiding that danger zone would reduce the likelihood of triggering major, irreversible climate changes, such as the disappearance of the Greenland ice cap. Two years ago the two of us provided a simple framework to relate future CO_2 emissions to this goal.

We contrasted two fifty-year futures. In one future, the emissions rate continues to grow at the pace of the past thirty years for the next fifty years, reaching 14 billion tons of carbon a year in 2056. (Higher or lower rates are, of course, plausible.) At that point, a tripling of preindustrial carbon concentrations would be very difficult to avoid, even with concerted efforts to decarbonize the world's energy systems over the following one hundred years. In the other future, emissions are frozen at the present value of 7 billion tons a year for the next fifty years and then reduced by about half over the following fifty years. In this way, a doubling of CO_2 levels can be avoided. The difference between these fifty-year emission paths — one ramping up and one flattening out — we called the stabilization triangle.

To hold global emissions constant while the world's economy continues to grow is a daunting task. Over the past thirty years, as the gross world product of goods and services grew at close to 3 percent a year on average, carbon emissions rose half as fast. Thus, the ratio of emissions to dollars of gross world product, known as the carbon intensity of the global economy, fell about 1.5 percent a year. For global emissions to be the same in 2056 as today, the carbon intensity will need to fall not half as fast but fully as fast as the global economy grows.

Two long-term trends are certain to continue and will help. First, as societies get richer, the services sector — education, health, leisure, banking, and so on — grows in importance relative to energy-intensive activities, such as steel production. All by itself, this shift lowers the carbon intensity of an economy.

Second, deeply ingrained in the patterns of technology evolution is the substitution of cleverness for energy. Hundreds of power plants are not needed today because the world has invested in much more efficient refrigerators, air conditioners, and motors than were available two decades ago. Hundreds of oil and gas fields have been developed more slowly because aircraft engines consume less fuel and the windows in gas-heated homes leak less heat.

The task of holding global emissions constant would be out of reach were it not for the fact that all the driving and flying in 2056 will be in vehicles not yet designed, most of the buildings that will be around then are not yet built, the locations of many of the communities that will contain these buildings and determine their inhabitants' commuting patterns have not yet been chosen, and utility owners are only now beginning to plan for the power plants that will be needed to light up those communities. Today's notoriously inefficient energy system can be replaced if the world gives unprecedented attention to energy efficiency. Dramatic changes are plausible over the next fifty years because so much of the energy canvas is still blank.

To make the task of reducing emissions vivid, we sliced the stabilization triangle into seven equal pieces, or "wedges," each representing 1 billion tons a year of averted emissions fifty years from now (starting from zero today). For example, a car driven 10,000 miles a year with a fuel efficiency of 30 miles per gallon (mpg) emits close to 1 ton of carbon annually. Transport experts predict that 2 billion cars will be zipping along the world's roads in 2056, each driven an average of 10,000 miles a year. If their average fuel efficiency were 30 mpg, their tailpipes would spew 2 billion tons of carbon that year. At 60 mpg, they would give off a billion tons. The latter scenario would therefore yield one wedge.

Wedges

In our framework, you are allowed to count as wedges only those differences in two 2056 worlds that result from deliberate carbon policy. The current pace of emissions growth already includes some steady reduction in carbon intensity. The goal is to reduce it even more. For instance, those who believe that cars will average 60 mpg in 2056 even in a world that pays no attention to carbon cannot count this improvement as a wedge, because it is already implicit in the baseline projection.

Moreover, you are allowed to count only strategies that involve the scaling up of technologies already commercialized somewhere in the world. You are not allowed to count pie in the sky. Our goal in developing the wedge framework was to be pragmatic and realistic — to propose engineering our way out of the problem and not waiting for the cavalry to come over the hill. We argued that even with these two counting rules, the world can fill all seven wedges, and in several different ways. Individual countries — operating within a framework of international cooperation — will decide which wedges to pursue, depending on their institutional and economic capacities, natural resource endowments, and political predilections.

To be sure, achieving nearly every one of the wedges requires new science and engineering to squeeze down costs and address the problems that inevitably accompany widespread deployment of new technologies. But holding CO_2 emissions in 2056 to their present rate, without choking off economic growth, is a desirable outcome within our grasp.

Ending the era of conventional coal-fired power plants is at the very top of the decarbonization agenda. Coal has become more competitive as a source of power and fuel because of energy security concerns and because of an increase in the cost of oil and gas. That is a problem, because a coal power plant burns twice as much carbon per unit of electricity as a natural gas plant. In the absence of a concern about carbon, the world's coal utilities could build a few thousand large (1,000-megawatt) conventional coal plants in the next fifty years. Seven hundred such plants emit one wedge's worth of carbon. Therefore, the world could take some big steps toward the target of freezing emissions by not building those plants. The time to start is now. Facilities built in this decade could easily be around in 2056.

Efficiency in electricity use is the most obvious substitute for coal. Of the 14 billion tons of carbon emissions projected for 2056, perhaps 6 billion will come from producing power, mostly from coal. Residential and commercial buildings account for 60 percent of global electricity demand today (70 percent in the U.S.) and will consume most of the new power. So cutting buildings' electricity use in half — by equipping them with superefficient lighting and appliances — could lead to two wedges. Another wedge would be achieved if industry finds additional ways to use electricity more efficiently.

Decarbonizing the Supply

Even after energy-efficient technology has penetrated deeply, the world will still need power plants. They can be coal plants, but they will need to be carbon-smart ones that capture the CO_2 and pump it into the ground. Today's high oil prices are lowering the cost of the transition to this technology, because captured CO_2 can often be sold to an oil company that injects it into oil fields to squeeze out more oil; thus, the higher the price of oil, the more valuable the captured CO_2. To achieve one wedge, utilities need to equip 800 large coal plants to capture and store nearly all the CO_2 otherwise emitted. Even in a carbon-constrained world, coal mining and coal power can stay in business, thanks to carbon capture and storage.

The large natural gas power plants operating in 2056 could capture and store their CO_2 too, perhaps accounting for yet another wedge. Renewable and nuclear energy can contribute as well. Renewable power can be produced from sunlight directly, either to energize photovoltaic cells or, using focusing mirrors, to heat a fluid and drive a turbine. Or the route can be indirect, harnessing hydropower and wind power, both of which rely on sun-driven weather patterns. The intermittency of renewable power does not diminish its capacity to contribute wedges; even if coal and natural gas plants provide the backup power, they run only part-time (in tandem with energy storage) and use less carbon than if they ran all year. Not strictly renewable, but also usually included in the family, is geothermal energy, obtained by mining the heat in the Earth's interior. Any of these sources, scaled up from its current contribution, could produce a wedge. One must be careful not to double-count the possibilities; the same coal plant can be left unbuilt only once.

Nuclear power is probably the most controversial of all the wedge strategies. IF the fleet of nuclear power plants were to expand by a factor of five by 2056, displacing conventional coal plants, it would provide two wedges. If the current fleet were to be shut down and replaced with modern coal plants without carbon capture and storage, the result would be *minus* one-half wedge. Whether nuclear power will be scaled up or down will depend on whether governments can find political solutions to waste disposal and on whether plants can run without accidents. (Nuclear plants are mutual hostages: the world's least well-run plants can imperil the future of all the others.) Also critical will be strict rules that prevent civilian nuclear technology from becoming a stimulus for nuclear weapons development. These rules will have to be uniform across all countries, so as to remove the sense of a double standard that has long been a spur to clandestine facilities.

Oil accounted for 43 percent of global carbon emissions from fossil fuels in 2002, while coal accounted for 27 percent; natural gas made up the remainder. More than half the oil was used for transport. So smartening up electricity production alone cannot fill the stabilization triangle; transportation too must be decarbonized. As with coal-fired electricity, at least a wedge may be available from each of three complementary options: reduced use, improved efficiency, and decarbonized energy sources. People can take fewer unwanted trips (telecommuting instead of vehicle commuting) and pursue the travel they cherish (adventure, family visits) in fuel-efficient vehicles running on low-carbon fuel. The fuel can be a product of crop residues or dedicated crops, hydrogen made from low-carbon electricity, or low-carbon electricity itself, charging an onboard battery. Sources of the low-carbon electricity could include wind, nuclear power, or coal with capture and storage.

Looming over this task is the prospect that in the interest of energy security, the transport system could become *more* carbon-intensive. That will happen if transport fuels are derived from coal instead of petroleum. Coal-based synthetic fuels, known as synfuels, provide a way to reduce global demand for oil, lowering its cost and decreasing global dependence on Middle East petroleum. But it is a decidedly climate-unfriendly strategy. A synfuel-powered car emits the same CO_2 as a gasoline-powered car, but synfuel fabrication from coal spews out far more carbon than does refining gasoline from crude oil — enough to double the emissions per mile of driving. From the perspective of mitigating climate change, it is fortunate that the emissions at a synfuels plant can be captured and stored. If business-as-usual trends did lead to the widespread adoption of synfuel, then capturing CO_2 at synfuels plants might well produce a wedge.

Not all wedges involve new energy technology. If all the farmers in the world practiced no-till agriculture rather than conventional plowing, they would contribute a wedge. Eliminating deforestation would result in two wedges, if the alternative were for deforestation to continue at current rates. Curtailing emissions of methane, which today contribute about half as much to greenhouse warming as CO_2, may provide more than one wedge: needed is a deeper understanding of the anaerobic biological emissions from cattle, rice paddies, and irrigated land. Lower birthrates can produce a wedge too — for example, if they hold the global population in 2056 near 8 billion people when it otherwise would have grown to 9 billion.

Action Plan

What set of policies will yield seven wedges? To be sure, the dramatic changes we anticipate in the fossil-fuel system, including routine use of CO_2 capture and storage, will require institutions that reliably communicate a price for present and future carbon emissions. We estimate that the price needed to jump-start this transition is in the ballpark of $100 to $200 per ton of carbon — the range that would make it cheaper for owners of coal plants to capture and store CO_2 than to vent it. The price might fall as technologies climb the learning curve. A carbon emissions price of $100 per ton is comparable to the current U.S. production credit for new renewable and nuclear energy relative to coal, and it is about half the current U.S. subsidy of ethanol relative to gasoline. It also was the price of CO_2 emissions in the European Union's emissions trading system for nearly a year, spanning 2005 and 2006. (One ton of carbon is carried in 3.7 tons of carbon dioxide, so this price is also $27 per ton of CO_2). Based on carbon content, $100 per ton of carbon is $12 per barrel of oil and $60 per ton of coal. It is 25 cents per gallon of gasoline and 2 cents per kilowatt-hour of electricity from coal.

But a price of CO_2 emissions on its own may not be enough. Governments may need to stimulate the commercialization of low-carbon technologies to increase the number of competitive options available in the future. Examples include wind, photovoltaic power, and hybrid cars. Also appropriate are policies designed to prevent the construction of long-lived capital facilities that are mismatched to future policy. Utilities, for instance, need to be encouraged to invest in CO_2 capture and storage for new coal power plants, which would be very costly to retrofit later. Still another set of policies can harness the capacity and energy of producers to promote efficiency — motivating power utilities to care about installation and maintenance of efficient appliances, natural gas companies to care about the buildings where their gas is burned, and oil companies to care about the engines that run on their fuel.

To freeze emissions at the current level, if one category of emissions goes up, another must come own. If emissions from natural gas increase, the combined emissions from oil and coal must decrease. If emissions from air travel climb, those from some other economic sector must fall. And if today's poor countries are to emit more, today's richer countries must emit less.

How much less? It is easy to bracket the answer. Currently the industrial nations — the members of the Organization for Economic Cooperation and Development (OECD) — account for almost exactly half the planet's CO_2 emissions, and the developing countries plus the nations formerly part of the Soviet Union account for the other half. In a world of constant total carbon emissions, keeping the OECD's share at 50 percent seems impossible to justify in the face of the enormous pent-up demand for energy in the non-OECD countries, where more than 80 percent of the world's people live. On the other hand, the OECD member states must emit *some* carbon in 2056. Simple arithmetic indicates that to hold global emissions rates steady, non-OECD emissions cannot even double.

One intermediate value results if all OECD countries were to meet the emissions-reduction target for the U.K. that was articulated in 2003 by Prime Minister Tony Blair — namely, a 60 percent reduction by 2050, relative to recent levels. The non-OECD countries could then emit 60 percent more CO_2. On average, by midcentury they would have one half the per capita emissions of the OECD countries. The CO_2 output of every country, rich or poor today, would be well below what it is generally projected to be in the absence of climate policy. In the case of the U.S., it would be about four times less.

Blair's goal would leave the average American emitting twice as much as the world average, as opposed to five times as much today. The U.S. could meet this goal in many ways. These strategies will be followed by most other countries as well. The resultant cross-pollination will lower every country's costs.

Fortunately, the goal of decarbonization does not conflict with the goal of eliminating the world's most extreme poverty. The extra carbon emissions produced when the world's nations accelerate the delivery of electricity and modern cooking fuel to the Earth's poorest people can be compensated for by, at most, one fifth of a wedge of emissions reductions elsewhere.

Beyond 2006

The stabilization triangle deals only with the first fifty-year leg of the future. One can imagine a relay race made of fifty-year segments, in which the first runner passes a baton to the second in 2056. Intergenerational equity requires that the two runners have roughly equally difficult tasks. It seems to us that the task we have given the second runner (to cut the 2056 emissions rate in half between 2056 and 2106) will not be harder than the task of the first runner (to keep global emissions in 2056 at present levels) — provided that between now and 2056 the world invests in research and development to get ready. A vigorous effort can prepare the revolutionary technologies that will give the second half of the century a running start. Those operations could include scrubbing CO_2 directly from the air, carbon storage in minerals, nuclear fusion, nuclear thermal hydrogen, and artificial photosynthesis. Conceivably, one or more of these technologies may arrive in time to help the first runner, although, as we have argued, the world should not count on it.

As we look back from 2056, if global emissions of CO_2 are indeed no larger than today's, what will have been accomplished? The world will have confronted energy production and energy efficiency at the consumer level, in all economic sectors and in economies at all levels of development. Buildings and lights and refrigerators, cars and trucks and planes, will be transformed. Transformed also will be the ways we use them.

The world will have a fossil-fuel energy system about as large as today's but one that is infused with modern controls and advanced materials and that is almost unrecognizably cleaner. There will be integrated production of power, fuels, and heat; greatly reduced air and water pollution; and extensive carbon capture and storage. Alongside the fossil energy system will be a nonfossil energy system approximately as large. Extensive direct and indirect harvesting of renewable energy will have brought about the revitalization of rural areas and the reclamation of degraded lands. If nuclear power is playing a large role, strong international enforcement mechanisms will have come into being to control the spread of nuclear technology from energy to weapons. Economic growth will have been maintained; the poor and the rich will both be richer. And our descendants will not be forced to exhaust so much treasure, innovation, and energy to ward off rising sea level, heat, hurricanes, and drought.

Critically, a planetary consciousness will have grown. Humanity will have learned to address its collective destiny — and to share the planet.

NEIL DEGRASSE TYSON

Delusions of Space Enthusiasts

FROM *Natural History*

HUMAN INGENUITY seldom fails to improve on the fruits of human invention. Whatever may have dazzled everyone on its debut is almost guaranteed to be superseded and, someday, to look quaint.

In 2000 B.C. a pair of ice skates made of polished animal bone and leather thongs was a transportation breakthrough. In 1610 Galileo's eight-power telescope was an astonishing tool of detection, capable of giving the senators of Venice a sneak peek at hostile ships before they could enter the lagoon. In 1887 the one-horsepower Benz Patent Motorwagen was the first commercially produced car powered by an internal combustion engine. In 1946 the 30-ton, showroom-sized ENIAC, with its 18,000 vacuum tubes and 6,000 manual switches, pioneered electronic computing. Today you can glide across roadways on in-line skates, gaze at images of faraway galaxies brought to you by the Hubble Space Telescope, cruise the autobahn in a 600-horsepower roadster, and carry your 3-pound laptop to an outdoor café.

Of course, such advances don't just fall from the sky. Clever people think them up. Problem is, to turn a clever idea into reality, somebody has to write the check. And when market forces shift, those somebodies may lose interest and the checks may stop coming. If computer companies had stopped innovating in 1978, your desk might still sport a 100-pound IBM 5110. If communications companies had stopped innovating in 1973, you might still be schlepping a 2-pound, 9-inch-long cell phone. And if in 1968 the U.S. space industry had stopped developing bigger and better rockets to launch humans beyond the moon, we'd never have surpassed the Saturn V rocket.

Oops!

Sorry about that. We haven't surpassed the Saturn V. The largest, most powerful rocket ever flown by anybody, ever, the 36-story-tall Saturn V was the first and only rocket to launch people from Earth to someplace else in the universe. It enabled every Apollo mission to the moon from 1969 through 1972, as well as the 1973 launch of Skylab 1, the first U.S. space station.

Inspired in part by the successes of the Saturn V and the momentum of the Apollo program, visionaries of the day foretold a future that never came to be: space habitats, moon bases, and Mars colonies up and running by the 1990s. But funding for the Saturn V evaporated as the moon missions wound down. Additional production runs were canceled, the manufacturers' specialized machine tools were destroyed, and skilled personnel had to find work on other projects. Today U.S. engineers can't even build a Saturn V clone.

What cultural forces froze the Saturn V rocket in time and space? What misconceptions led to the gap between expectation and reality?

Soothsaying tends to come in two flavors: doubt and delirium. It was doubt that led skeptics to declare that the atom would never be split, the sound barrier would never be broken, and people would never want or need computers in their homes. But in the case of the Saturn V rocket, it was delirium that misled futurists into assuming the Saturn V was an auspicious beginning — never considering that it could instead be an end.

On December 30, 1900, for its last Sunday paper of the nineteenth century, the *Brooklyn Daily Eagle* published a sixteen-page supplement headlined "THINGS WILL BE SO DIFFERENT A HUNDRED YEARS HENCE." The contributors — business leaders, military men, pastors, politicians, and experts of every persuasion — imagined what housework, poverty, religion, sanitation, and war would be like in the year 2000. They enthused about the potential of electricity and the automobile. There was even a map of the world-to-be, showing an American Federation comprising most of the Western Hemisphere from the lands above the Arctic Circle down to the archipelago of Tierra del Fuego — plus sub-Saharan Africa, the southern half of Australia, and all of New Zealand.

Most of the writers portrayed an expansive future. But not all. George H. Daniels, a man of authority at the New York Central and Hudson River Railroad, peered into his crystal ball and boneheadedly predicted:

> It is scarcely possible that the twentieth century will witness improvements in transportation that will be as great as were those of the nineteenth century.

Elsewhere in his article, Daniels envisioned affordable global tourism and the diffusion of white bread to China and Japan. Yet he simply couldn't imagine what might replace steam as the power source for ground transportation, let alone a vehicle moving through the air. Even though he stood on the doorstep of the twentieth century, this manager of the world's biggest railroad system could not see beyond the automobile, the locomotive, and the steamship.

Three years later, almost to the day, Wilbur and Orville Wright made the first-ever series of powered, controlled, heavier-than-air flights. By 1957 the USSR launched the first satellite into Earth orbit. And in 1969 two Americans became the first human beings to walk on the moon.

Daniels is hardly the only person to have misread the technological future. Even experts who aren't totally deluded can have tunnel vision. On page 13 of the *Eagle*'s Sunday supplement, the principal examiner at the U.S. Patent Office, W. W. Townsend, wrote, "The automobile may be the vehicle of the decade, but the air ship is the conveyance of the century." Sounds visionary, until you read further. What he was talking about were blimps and zeppelins. Both Daniels and Townsend, otherwise well-informed citizens of a changing world, were clueless about what tomorrow's technology would bring.

Even the Wrights were guilty of doubt about the future of aviation. In 1901, discouraged by a summer's worth of unsuccessful tests with a glider, Wilbur told Orville it would take another fifty years for someone to fly. Nope: the birth of aviation was just two years away. On the windy, chilly morning of December 17, 1903, starting from a North Carolina sand dune called Kill Devil Hill, Orville was the first to fly the brothers' 600-pound plane through the air. His epochal journey lasted twelve seconds and covered 120 feet — a distance just shy of the wingspan of a Boeing 757.

Judging by what the mathematician, astronomer, and Royal Society gold medalist Simon Newcomb had published just two months earlier, the flights from Kill Devil Hill should never have taken place when they did:

> Quite likely the twentieth century is destined to see the natural forces which will enable us to fly from continent to continent with a speed far exceeding that of the bird.
>
> But when we inquire whether aerial flight is possible in the present state of our knowledge; whether, with such materials as we possess, a combination of steel, cloth and wire can be made which, moved by the power of electricity or steam, shall form a successful flying machine, the outlook may be altogether different.

Some representatives of informed public opinion went even further. The *New York Times* was steeped in doubt just one week before the Wright brothers went aloft in the original *Wright Flyer.* Writing on December 10, 1903 — not about the Wrights but about their illustrious and publicly funded competitor, Samuel P. Langley, an astronomer, physicist, and chief administrator of the Smithsonian Institution — the *Times* declared:

> We hope that Professor Langley will not put his substantial greatness as a scientist in further peril by continuing to waste his time, and the money involved, in further airship experiments. Life is short, and he is capable of services to humanity incomparably greater than can be expected to result from trying to fly.

You might think attitudes would have changed as soon as people from several countries had made their first flights. But no. Wilbur Wright wrote in 1909 that no flying machine would ever make the journey from New York to Paris. Richard Burdon Haldane, the British secretary of war, told Parliament in 1909 that even though the airplane might one day be capable of great things, "from the war point of view, it is not so at present." Ferdinand Foch, a highly regarded French military strategist and the supreme commander of the Allied forces near the end of the First World War, opined in 1911 that airplanes were interesting toys but had no military value.

Late that same year, near Tripoli, an Italian plane became the first to drop a bomb.

Early attitudes about flight beyond Earth's atmosphere followed a similar trajectory. True, plenty of philosophers, scientists, and sci-fi writers had thought long and hard about outer space. The sixteenth-century philosopher-friar Giordano Bruno proposed that intelligent beings inhabited an infinitude of worlds. The seventeenth-century soldier-writer Savinien de Cyrano de Bergerac portrayed the moon as a world with forests, violets, and people.

But those writings were fantasies, not blueprints for action. By the early twentieth century, electricity, telephones, automobiles, radios, airplanes, and countless other engineering marvels were all becoming basic features of modern life. So couldn't earthlings build machines capable of space travel? Many people who should have known better said it couldn't be done, even after the successful 1942 test launch of the world's first long-range ballistic missile: Germany's deadly V-2 rocket. Capable of punching through Earth's atmosphere, it was a crucial step toward reaching the moon.

Richard van der Riet Woolley, the eleventh British Astronomer Royal, is the source of a particularly woolly remark. When he landed in London after a thirty-six-hour flight from Australia, some reporters asked him about space travel. "It's utter bilge," he answered. That was in early 1956. In early 1957 Lee De Forest, a prolific American inventor who helped birth the age of electronics, declared, "Man will never reach the moon, regardless of all future scientific advances." Remember what happened in late 1957? Not just one but two Soviet *Sputnik*s entered Earth's orbit. The space race had begun.

Whenever someone says an idea is "bilge" (British for "baloney"), you must first ask whether it violates any well-tested laws of physics. If so, the idea is likely to be bilge. If not, the only challenge is to find a clever engineer — and, of course, a committed source of funding.

The day the Soviet Union launched *Sputnik 1,* a chapter of science fiction became science fact, and the future became the present. All of a sudden, futurists went overboard with their enthusiasm. The delusion that technology would advance at lightning speed replaced the delusion that it would barely advance at all. Experts went from having much too little confidence in the pace of technology to having much too much. And the guiltiest people of all were the space enthusiasts.

Commentators became fond of twenty-year intervals, within which some previously inconceivable goal would supposedly be accomplished. On January 6, 1967, in a front-page story, the *Wall Street Journal* announced: "The most ambitious U.S. space endeavor in the years ahead will be the campaign to land men on neighboring Mars. Most experts estimate the task can be accomplished by 1985." The very next month, in its debut issue, *The Futurist* magazine announced that according to long-range forecasts by the RAND Corporation, a pioneer think-tank, there was a 60 percent probability that a manned lunar base would exist by 1986. In *The Book of Predictions,* published in 1980, the rocket pioneer Robert C. Truax forecast that 50,000 people would be living and working in space by the year 2000. When that benchmark year arrived, people were indeed living and working in space. But the tally was not 50,000. It was three: the first crew of the International Space Station.

All those visionaries (and countless others) never really grasped the forces that drive technological progress. In Wilbur and Orville's day, you could tinker your way into major engineering advances. Their first airplane did not require a grant from the National Science Foundation: they funded it through their bicycle business. The brothers constructed the wings and fuselage themselves, with tools they already owned, and got their resourceful bicycle mechanic, Charles E. Taylor, to design and hand-build the engine. The operation was basically two guys and a garage.

Space exploration unfolds on an entirely different scale. The first moonwalkers were two guys too — Neil Armstrong and Buzz Aldrin — but behind them loomed the force of a mandate from an assassinated president, 10,000 engineers, $100 billion, and a Saturn V rocket.

Notwithstanding the sanitized memories so many of us have of the Apollo era, Americans were not first on the moon because we're explorers by nature or because our country is committed to the pursuit of knowledge. We got to the moon first because the United States was out to beat the Soviet Union, to win the cold war any way we could. John F. Kennedy made that clear when he complained to top NASA officials in November 1962:

> I'm not that interested in space. I think it's good, I think we ought to know about it, we're ready to spend reasonable amounts of money. But we're talking about these fantastic expenditures which wreck our budget and all these other domestic programs and the only justification for it in my opinion to do it in this time or fashion is because we hope to beat them [the Soviet Union] and demonstrate that starting behind, as we did by a couple of years, by God, we passed them.

Like it or not, war (cold or hot) is the most powerful funding driver in the public arsenal. When a country wages war, money flows like floodwaters. Lofty goals — such as curiosity, discovery, exploration, and science — can get you money for modest-sized projects, provided they resonate with the political and cultural views of the moment. But big, expensive activities are inherently long-term, and require sustained investment that must survive economic fluctuations and changes in the political winds.

In all eras, across time and culture, only three drivers have fulfilled that funding requirement: war, greed, and the celebration of royal or religious power. The Great Wall of China, the pyramids of Egypt, the Gothic cathedrals of Europe, the U.S. interstate highway system, the voyages of Columbus and Cook — nearly every major undertaking owes its existence to one or more of those three drivers. Today, as the power of kings is supplanted by elected governments and the power of religion is often expressed in nonarchitectural undertakings, that third driver has lost much of its sway, leaving war and greed to run the show. Sometimes those two drivers work hand in hand, as in the art of profiteering from the art of war. But war itself remains the ultimate and most compelling rationale.

Having been born the same week NASA was founded, I was eleven years old during the voyage of *Apollo 11,* and had already identified the universe as my life's passion. Unlike so many other people who watched Neil Armstrong's first steps on the moon, I wasn't jubilant. I was simply relieved that someone was finally exploring another world. To me, *Apollo 11* was clearly the beginning of an era.

But I too was delirious. The lunar landings continued for three and a half years. Then they stopped. The Apollo program became the end of an era, not the beginning. And as the moon voyages receded in time and memory, they seemed ever more unreal in the history of human projects.

Unlike the first ice skates or the first airplane or the first desktop computer — artifacts that make us all chuckle when we see them today — the first rocket to the moon, the 364-foot-tall Saturn V, elicits awe, even reverence. Three Saturn V relics lie in state at the Johnson Space Center in Texas, the Kennedy Space Center in Florida, and the U.S. Space and Rocket Center in Alabama. Streams of worshippers walk the length of each rocket. They touch the mighty rocket nozzles at the base and wonder how something so large could ever have bested Earth's gravity. To transform their awe into chuckles, our country will have to resume the effort to "boldly go where no man has gone before." Only then will the Saturn V look as quaint as every other invention that human ingenuity has paid the compliment of improving upon.

ETHAN WATTERS

DNA Is Not Destiny

FROM *Discover*

BACK IN 2000, Randy Jirtle, a professor of radiation oncology at Duke University, and his postdoctoral student Robert Waterland designed a groundbreaking genetic experiment that was simplicity itself. They started with pairs of fat yellow mice known to scientists as agouti mice, so called because they carry a particular gene — the agouti gene — that in addition to making the rodents ravenous and yellow renders them prone to cancer and diabetes. Jirtle and Waterland set about to see if they could change the unfortunate genetic legacy of these little creatures.

Typically, when agouti mice breed, most of the offspring are identical to the parents: just as yellow, fat as pincushions, and susceptible to life-shortening disease. The parent mice in Jirtle and Waterland's experiment, however, produced a majority of offspring that looked altogether different. These young mice were slender and mousy brown. Moreover, they did not display their parents' susceptibility to cancer and diabetes and lived to a spry old age. The effects of the agouti gene had been virtually erased.

Remarkably, the researchers effected this transformation without altering a single letter of the mouse's DNA. Their approach instead was radically straightforward — they changed the moms' diet. Starting just before conception, Jirtle and Waterland fed a test group of mother mice a diet rich in methyl donors, small chemical clusters that can attach to a gene and turn it off. These molecules are common in the environment and are found in many foods, including onions, garlic, and beets, and in the food supplements often given to pregnant women. After being consumed by the mothers, the methyl donors worked their way into the developing embryos' chromosomes and onto the critical agouti gene. The mothers passed along the agouti gene to their children intact, but thanks to their methyl-rich pregnancy diet, they had added to the gene a chemical switch that dimmed the gene's deleterious effects.

"It was a little eerie and a little scary to see how something as subtle as a nutritional change in the pregnant mother rat could have such a dramatic impact on the gene expression of the baby," Jirtle says. "The results showed how important epigenetic changes could be."

Our DNA — specifically the 25,000 genes identified by the Human Genome Project — is now widely regarded as the instruction book for the human body. But genes themselves need instructions for what to do, and where and when to do it. A human liver cell contains the same DNA as a brain cell, yet somehow it knows to code only those proteins needed for the functioning of the liver. Those instructions are found not in the letters of the DNA itself but on it, in an array of chemical markers and switches, known collectively as the epigenome, that lie along the length of the double helix. These epigenetic switches and markers in turn help switch on or off the expression of particular genes. Think of the epigenome as a complex software code, capable of inducing the DNA hardware to manufacture an impressive variety of proteins, cell types, and individuals.

In recent years, epigenetics researchers have made great strides in understanding the many molecular sequences and patterns that determine which genes can be turned on and off. Their work has made it increasingly clear that for all the popular attention devoted to genome-sequencing projects, the epigenome is just as critical as DNA to the healthy development of organisms, humans included. Jirtle and Waterland's experiment was a benchmark demonstration that the epigenome is sensitive to cues from the environment. More and more, researchers are finding that an extra bit of a vitamin, a brief exposure to a toxin, even an added dose of mothering can tweak the epigenome — and thereby alter the software of our genes — in ways that affect an individual's body and brain for life.

The even greater surprise is the recent discovery that epigenetic signals from the environment can be passed on from one generation to the next, sometimes for several generations, without changing a single gene sequence. It's well established, of course, that environmental effects like radiation, which alter the genetic sequences in a sex cell's DNA, can leave a mark on subsequent generations. Likewise, it's known that the environment in a mother's womb can alter the development of a fetus. What's eye-opening is a growing body of evidence suggesting that the epigenetic changes wrought by one's diet, behavior, or surroundings can work their way into the germ line and echo far into the future. Put simply, and as bizarre as it may sound, what you eat or smoke today could affect the health and behavior of your great-grandchildren.

All of these discoveries are shaking the modern biological and social certainties about genetics and identity. We commonly accept the notion that through our DNA we are destined to have particular body shapes, personalities, and diseases. Some scholars even contend that the genetic code predetermines intelligence and is the root cause of many social ills, including poverty, crime, and violence. "Gene as fate" has become conventional wisdom. Through the study of epigenetics, that notion at last may be proved outdated. Suddenly, for better or worse, we appear to have a measure of control over our genetic legacy.

"Epigenetics is proving we have some responsibility for the integrity of our genome," Jirtle says. "Before, genes predetermined outcomes. Now everything we do — everything we eat or smoke — can affect our gene expression and that of future generations. Epigenetics introduces the concept of free will into our idea of genetics."

Scientists are still coming to understand the many ways that epigenetic changes unfold at the biochemical level. One form of epigenetic change physically blocks access to the genes by altering what is called the histone code. The DNA in every cell is tightly wound around proteins known as histones and must be unwound to be transcribed. Alterations to this packaging cause certain genes to be more or less available to the cell's chemical machinery and so determine whether those genes are expressed or silenced. A second, well-understood form of epigenetic signaling, called DNA methylation, involves the addition of a methyl group — a carbon atom plus three hydrogen atoms — to particular bases in the DNA sequence. This interferes with the chemical signals that would put the gene into action and thus effectively silences the gene.

Until recently, the pattern of an individual's epigenome was thought to be firmly established during early fetal development. Although that is still seen as a critical period, scientists have lately discovered that the epigenome can change in response to the environment throughout an individual's lifetime.

"People used to think that once your epigenetic code was laid down in early development, that was it for life," says Moshe Szyf, a pharmacologist with a bustling lab at McGill University in Montreal. "But life is changing all the time, and the epigenetic code that controls your DNA is turning out to be the mechanism through which we change along with it. Epigenetics tells us that little things in life can have an effect of great magnitude."

Szyf has been a pioneer in linking epigenetic changes to the development of diseases. He long ago championed the idea that epigenetic patterns can shift through life and that those changes are important in the establishment and spread of cancer. For fifteen years, however, he had little luck convincing his colleagues. One of his papers was dismissed by a reviewer as a "misguided attempt at scientific humor." On another occasion, a prominent scientist took him aside and told him bluntly, "Let me be clear: cancer is genetic in origin, not epigenetic."

Despite such opposition, Szyf and other researchers have persevered. Through numerous studies, Szyf has found that common signaling pathways known to lead to cancerous tumors also activate the DNA-methylation machinery; knocking out one of the enzymes in that pathway prevents the tumors from developing. When genes that typically act to suppress tumors are methylated, the tumors metastasize. Likewise, when genes that typically promote tumor growth are demethylated — that is, the dimmer switches that are normally present are removed — those genes kick into action and cause tumors to grow.

Szyf is now far from alone in the field. Other researchers have identified dozens of genes, all related to the growth and spread of cancer, that become over- or undermethylated when the disease gets under way. The bacterium *Helicobacter*, believed to be a cause of stomach cancer, has been shown to trigger potentially cancer-inducing epigenetic changes in gut cells. Abnormal methylation patterns have been found in many cancers of the colon, stomach, cervix, prostate, thyroid, and breast.

Szyf views the link between epigenetics and cancer with a hopeful eye. Unlike genetic mutations, epigenetic changes are potentially reversible. A mutated gene is unlikely to mutate back to normal; the only recourse is to kill or cut out all the cells carrying the defective code. But a gene with a defective methylation pattern might very well be encouraged to reestablish a healthy pattern and continue to function. Already one epigenetic drug, 5-azacytidine, has been approved by the Food and Drug Administration for use against myelodysplastic syndrome, also known as preleukemia or smoldering leukemia. At least eight other epigenetic drugs are currently in different stages of development or human trials.

Methylation patterns also hold promise as diagnostic tools, potentially yielding critical information about the odds that a cancer will respond to treatment. A Berlin-based company called Epigenomics, in partnership with Roche Pharmaceuticals, expects to bring an epigenetic screening test for colon cancer to market by 2008. They are working on similar diagnostic tools for breast cancer and prostate cancer. Szyf has cofounded a company, MethylGene, that so far has developed two epigenetic cancer drugs with promising results in human trials. Others have published data on animal subjects suggesting an epigenetic component to inflammatory diseases like rheumatoid arthritis, neurodegenerative diseases, and diabetes.

Other researchers are focusing on how people might maintain the integrity of their epigenomes through diet. Baylor College of Medicine obstetrician and geneticist Ignatia Van den Veyver suggests that once we understand the connection between our epigenome and diseases like cancer, lifelong "methylation diets" may be the trick to staying healthy. Such diets, she says, could be tailored to an individual's genetic makeup, as well as to his or her exposure to toxins or cancer-causing agents.

In 2003 biologist Ming Zhu Fang and her colleagues at Rutgers University published a paper in the journal *Cancer Research* on the epigenetic effects of green tea. In animal studies, green tea prevented the growth of cancers in several organs. Fang found that epigallocatechin-3-gallate (EGCG), the major polyphenol from green tea, can prevent deleterious methylation dimmer switches from landing on (and shutting down) certain cancer-fighting genes. The researchers described the study as the first to demonstrate that a consumer product can inhibit DNA methylation. Fang and her colleagues have since gone on to show that genistein and other compounds in soy show similar epigenetic effects.

Meanwhile, epigenetic researchers around the globe are rallying behind the idea of a human epigenome project, which would aim to map our entire epigenome. The Human Genome Project, which sequenced the 3 billion pairs of nucleotide bases in human DNA, was a piece of cake in comparison: epigenetic markers and patterns are different in every tissue type in the human body and also change over time. "The epigenome project is much more difficult than the Human Genome Project," Jirtle says. "A single individual doesn't have one epigenome but a multitude of them."

Research centers in Japan, Europe, and the United States have all begun individual pilot studies to assess the difficulty of such a project. The early signs are encouraging. In June the European Human Epigenome Project released its data on epigenetic patterns of three human chromosomes. A recent flurry of conferences have forwarded the idea of creating an international epigenome project that could centralize the data, set goals for different groups, and standardize the technology for decoding epigenetic patterns.

Until recently, the idea that your environment might change your heredity without changing a gene sequence was scientific heresy. Everyday influences — the weights Dad lifts to make himself muscle-bound, the diet regimen Mom follows to lose pounds — don't produce stronger or slimmer progeny, because those changes don't affect the germ cells involved in making children. Even after the principles of epigenetics came to light, it was believed that methylation marks and other epigenetic changes to a parent's DNA were lost during the process of cell division that generates eggs and sperm and that only the gene sequence remained. In effect, it was thought, germ cells wiped the slate clean for the next generation.

That turns out not to be the case. In 1999 biologist Emma Whitelaw, now at the Queensland Institute of Medical Research in Australia, demonstrated that epigenetic marks could be passed from one generation of mammals to the next. (The phenomenon had already been demonstrated in plants and yeast.) Like Jirtle and Waterland in 2003, Whitelaw focused on the agouti gene in mice, but the implications of her experiment span the animal kingdom.

"It changes the way we think about information transfer across generations," Whitelaw says. "The mindset at the moment is that the information we inherit from our parents is in the form of DNA. Our experiment demonstrates that it's more than just DNA you inherit. In a sense that's obvious, because what we inherit from our parents are chromosomes, and chromosomes are only fifty percent DNA. The other fifty percent is made up of protein molecules, and these proteins carry the epigenetic marks and information."

Michael Meaney, a biologist at McGill University and a frequent collaborator with Szyf, has pursued an equally provocative notion: that some epigenetic changes can be induced after birth, through a mother's physical behavior toward her newborn. For years Meaney sought to explain some curious results he had observed involving the nurturing behavior of rats. Working with graduate student Ian Weaver, Meaney compared two types of mother rats: those that patiently licked their offspring after birth and those that neglected their newborns. The licked newborns grew up to be relatively brave and calm (for rats). The neglected newborns grew into the sort of rodents that nervously skitter into the darkest corner when placed in a new environment.

Traditionally, researchers might have offered an explanation on one side or the other of the nature-versus-nurture divide. Either the newborns inherited a genetic propensity to be skittish or brave (nature), or they were learning the behavior from their mothers (nurture). Meaney and Weaver's results didn't fall neatly into either camp. After analyzing the brain tissue of both licked and nonlicked rats, the researchers found distinct differences in the DNA methylation patterns in the hippocampus cells of each group. Remarkably, the mother's licking activity had the effect of removing dimmer switches on a gene that shapes stress receptors in the pup's growing brain. The well-licked rats had better-developed hippocampi and released less of the stress hormone cortisol, making them calmer when startled. In contrast, the neglected pups released much more cortisol, had less developed hippocampi, and reacted nervously when startled or in new surroundings. Through a simple maternal behavior, these mother rats were literally shaping the brains of their offspring.

How exactly does the mother's behavior cause the epigenetic change in her pup? Licking and grooming release serotonin in the pup's brain, which activates serotonin receptors in the hippocampus. These receptors send proteins called transcription factors to turn on the gene that inhibits stress responses. Meaney, Weaver, and Szyf think that the transcription factors, which normally regulate genes in passing, also carry methylation machinery that can alter gene expression permanently. In two subsequent studies, Meaney and his colleagues were even able to reverse the epigenetic signals by injecting the drug trichostatin A into the brains of adult rats. In effect, they were able to simulate the effect of good (and bad) parenting with a pharmaceutical intervention. Trichostatin, interestingly, is chemically similar to the drug valproate, which is used clinically in people as a mood stabilizer.

Meaney says the link between nurturing and brain development is more than just a curious cause and effect. He suggests that making postnatal changes to an offspring's epigenome offers an adaptive advantage. Through such tweaking, mother rats have a last chance to mold their progeny to suit the environment they were born into. "These experiments emphasize the importance of context on the development of a creature," Meaney says. "They challenge the overriding theories of both biology and psychology. Rudimentary adaptive responses are not innate or passively emerging from the genome but are molded by the environment."

Meaney now aims to see whether similar epigenetic changes occur when human mothers caress and hold their infants. He notes that the genetic sequence silenced by attentive mother rats has a close parallel in the human genome, so he expects to find a similar epigenetic influence. "It's just not going to make any sense if we don't find this in humans as well. The story is going to be more complex than with the rats because we'll have to take into account more social influences, but I'm convinced we're going to find a connection."

In an early study, which provided circumstantial evidence, Meaney examined magnetic resonance imaging brain scans of adults who began life as low-birth-weight babies. Those adults who reported in a questionnaire that they had a poor relationship with their mother were found to have hippocampi that were significantly smaller than average. Those adults who reported having had a close relationship with their mother, however, showed perfectly normal-sized hippocampi. Meaney acknowledges the unreliability of subjects reporting on their own parental relationships; nonetheless, he strongly suspects that the quality of parenting was responsible for the different shapes of the brains of these two groups.

In an effort to solidify the connection, he and other researchers have launched an ambitious five-year multimillion-dollar study to examine the effects of early nurturing on hundreds of human babies. As a test group, he's using severely depressed mothers, who often have difficulty bonding with and caring for their newborns and as a result tend to caress their babies less than mothers who don't experience depression or anxiety. The question is whether the babies of depressed mothers show the distinct brain shapes and patterns indicative of epigenetic differences.

The science of epigenetics opens a window onto the inner workings of many human diseases. It also raises some provocative new questions. Even as we consider manipulating the human epigenome to benefit our health, some researchers are concerned that we may already be altering our epigenomes unintentionally, and perhaps not for the better. Jirtle notes that the prenatal vitamins that physicians commonly encourage pregnant women to take to reduce the incidence of birth defects in their infants include some of the same chemicals that Jirtle fed to his agouti mice. In effect, Jirtle wonders whether his mouse experiment is being carried out wholesale on American women.

"On top of the prenatal vitamins, every bit of grain product that we eat in the country is now fortified with folic acid," Jirtle notes, and folic acid is a known methyl donor. "In addition, some women take multivitamins that also have these compounds. They're getting a triple hit."

While the prenatal supplements have an undisputed positive effect, Jirtle says, no one knows where else in the fetal genome those gene-silencing methyl donors might be landing. A methyl tag that has a positive effect on one gene might have a deleterious effect if it happens to fall somewhere else. "It's the American way to think, 'If a little is good, a lot is great.' But that is not necessarily the case here. You might be overmethylating certain genes, which could potentially cause other things like autism and other negative outcomes."

Szyf shares the concern. "Fueling the methylation machinery

through dietary supplements is a dangerous experiment, because there is likely to be a plethora of effects throughout a lifetime." In the future, he believes, epidemiologists will have their hands full looking for possible epigenetic consequences of these public-health choices. "Did this change in diet increase cancer risk? Did it increase depression? Did it increase schizophrenia? Did it increase dementia or Alzheimer's? We don't know yet. And it will take some time to sort it out."

The implications of the epigenetic revolution are even more profound in light of recent evidence that epigenetic changes made in the parent generation can turn up not just one but several generations down the line, long after the original trigger for change has been removed. In 2004, Michael Skinner, a geneticist at Washington State University, accidentally discovered an epigenetic effect in rats that lasts at least four generations. Skinner was studying how a commonly used agricultural fungicide, when introduced to pregnant mother rats, affected the development of the testes of fetal rats. He was not surprised to discover that male rats exposed to high doses of the chemical while in utero had lower sperm counts later in life. The surprise came when he tested the male rats in subsequent generations — the grandsons of the exposed mothers. Although the pesticide had not changed one letter of their DNA, these second-generation offspring also had low sperm counts. The same was true of the next generation (the great-grandsons) and the next.

Such results hint at a seemingly anti-Darwinian aspect of heredity. Through epigenetic alterations, our genomes retain something like a memory of the environmental signals received during the lifetimes of our parents, grandparents, great-grandparents, and perhaps even more distant ancestors. So far, the definitive studies have involved only rodents. But researchers are turning up evidence suggesting that epigenetic inheritance may be at work in humans as well.

In November 2005, Marcus Pembrey, a clinical geneticist at the Institute of Child Health in London, attended a conference at Duke University to present intriguing data drawn from two centuries of records on crop yields and food prices in an isolated town in northern Sweden. Pembrey and Swedish researcher Lars Olov Bygren noted that fluctuations in the town's food supply may have health effects spanning at least two generations. Grandfathers who lived their preteen years during times of plenty were more likely to have grandsons with diabetes — an ailment that doubled the grandsons' risk of early death. Equally notable was that the effects were sex-specific. A grandfather's access to a plentiful food supply affected the mortality rates of his grandsons only, not those of his granddaughters, and a paternal grandmother's experience of feast affected the mortality rates of her granddaughters, not her grandsons.

This led Pembrey to suspect that genes on the sex-specific X and Y chromosomes were being affected by epigenetic signals. Further analysis supported his hunch and offered insight into the signaling process. It turned out that timing — the ages at which grandmothers and grandfathers experienced a food surplus — was critical to the intergenerational impact. The granddaughters most affected were those whose grandmothers experienced times of plenty while in utero or as infants, precisely the time when the grandmothers' eggs were forming. The grandsons most affected were those whose grandfathers experienced plenitude during the so-called slow growth period, just before adolescence, which is a key stage for the development of sperm.

The studies by Pembrey and other epigenetics researchers suggest that our diet, behavior, and environmental surroundings today could have a far greater impact than imagined on the health of our distant descendants. "Our study has shown a new area of research that could potentially make a major contribution to public health and have a big impact on the way we view our responsibilities toward future generations," Pembrey says.

The logic applies backward as well as forward: some of the disease patterns prevalent today may have deep epigenetic roots. Pembrey and several other researchers, for instance, have wondered whether the current epidemic of obesity, commonly blamed on the excesses of the current generation, may partially reflect lifestyles adopted by our forebears two or more generations back.

Michael Meaney, who studies the impact of nurturing, likewise wonders what the implications of epigenetics are for social policy. He notes that early child-parent bonding is made more difficult by the effects of poverty, dislocation, and social strife. Those factors can certainly affect the cognitive development of the children directly involved. Might they also affect the development of future generations through epigenetic signaling?

"These ideas are likely to have profound consequences when you start to talk about how the structure of society influences cognitive development," Meaney says. "We're beginning to draw cause-and-effect arrows between social and economic macrovariables down to the level of the child's brain. That connection is potentially quite powerful."

Lawrence Harper, a psychologist at the University of California at Davis, suggests that a wide array of personality traits, including temperament and intelligence, may be affected by epigenetic inheritance. "If you have a generation of poor people who suffer from bad nutrition, it may take two or three generations for that population to recover from that hardship and reach its full potential," Harper says. "Because of epigenetic inheritance, it may take several generations to turn around the impact of poverty or war or dislocation on a population."

Historically, genetics has not meshed well with discussions of social policy; it's all too easy to view disadvantaged groups — criminals, the poor, the ethnically marginalized — as somehow fated by DNA to their condition. The advent of epigenetics offers a new twist and perhaps an opportunity to understand with more nuance how nature and nurture combine to shape the society we live in today and hope to live in tomorrow.

"Epigenetics will have a dramatic impact on how we understand history, sociology, and political science," says Szyf. "If environment has a role to play in changing your genome, then we've bridged the gap between social processes and biological processes. That will change the way we look at everything."

Contributors' Notes

Other Notable Science and Nature Writing of 2006

Contributors' Notes

Paul Bennett contributes regularly to *National Geographic* and other magazines and is a European correspondent for *Architectural Record.* He lives in Paris with his wife and two daughters.

Emily Case teaches middle and high school science at a public school in western Massachusetts, a job which allows her to explore the wonder of science and the power of inquiry with young people. She has a B.A. in biology from Vassar College and an M.Ed. from the University of Massachusetts, Amherst.

Susan Casey is the author of the *New York Times* bestseller *The Devil's Teeth: A True Story of Obsession and Survival Among America's Great White Sharks* (2005). *The Devil's Teeth,* Casey's first book, was also a *San Francisco Chronicle* bestseller, a Book Sense bestseller, a Barnes & Noble Discover Selection, a Library Journal Best Book of 2005, a 2005 NPR Summer Reading Selection, and a Hudson News Best Book of 2006. From 2003 to 2006, Casey was the development editor of Time Inc., where she had previously been the editor of *Sports Illustrated Women* and an editor at large. She also served as the creative director of *Outside* magazine. Her writing has appeared in *Sports Illustrated, Esquire, Time, Fortune,* the *New York Times, Best Life,* and *Outside.* A native of Toronto, Casey lives in New York City.

Richard Conniff has collected tarantulas in the Peruvian Amazon, tracked wild dogs in Botswana's Okavango Delta, climbed the Mountains of the Moon in western Uganda, and trekked through the Himalayas of Bhutan in pursuit of tigers and the mythical migur. His latest book is *The Ape in the Corner Office: How to Make Friends, Win Fights, and Work Smarter by Under-*

standing Human Nature. His magazine work on invertebrates won the 1997 National Magazine Award and was later collected in his book *Spineless Wonders: Strange Tales from the Invertebrate World.* Conniff is also an occasional commentator on NPR's *All Things Considered,* and he has written or presented television shows for National Geographic, Discovery Channel, and the BBC. With the help of a 2007 Guggenheim Foundation Fellowship, he is currently at work on a project about the discovery of new species.

Alison Hawthorne Deming is the author of three books of poems, most recently *Genius Loci* (2005), and three books of nonfiction, most recently *Writing the Sacred into the Real* (2001). She is a professor of creative writing at the University of Arizona in Tucson.

Brian Doyle is the editor of *Portland Magazine* at the University of Portland and the author of eight books of essays, nonfiction, and "proems," most recently a collection called *Epiphanies & Elegies.*

Helen Fields has been a reporter at *U.S. News & World Report* and is now an editor for *National Geographic.*

Patricia Gadsby has worked as a senior editor for *Discover* and continues to contribute to the magazine, writing articles that illuminate the everyday acts of eating and cooking with a gleam of science. She has written about the sense of taste, the chemistry of "fishy" smells, the paradox of the old Inuit diet, the microorganisms in sourdough bread, and the genetic history of chocolate. She lives with her scientist husband in New York and in Woods Hole, Massachusetts. She enjoys fishing and cooking the catch.

James Gleick is an author, reporter, and essayist. His most recent book, *Isaac Newton,* was a Pulitzer Prize finalist in 2003 and a national bestseller, as were *Chaos: Making a New Science* (1987) and *Genius: The Life and Science of Richard Feynman* (1992).

John Horgan is a science journalist and director of the Center for Science Writings at the Stevens Institute of Technology, in Hoboken, New Jersey. A former senior writer at *Scientific American* (1986–1997), he has also written for the *New York Times, Time, Newsweek,* the *Washington Post,* the *Los Angeles Times,* the *New Republic, Slate, Discover,* the London *Times,* the *Times Literary Supplement, New Scientist,* and other publications around the world. His three books include *Rational Mysticism* (2003), *The Undiscovered Mind* (1999), and *The End of Science* (1996).

William Langewiesche is the international correspondent for *Vanity Fair.* He is the author of five books, including *American Ground: Unbuilding the World Trade Center,* an insider's account of the nine-month cleanup of the Twin Towers. He has been nominated for several writing and journalism prizes, such as the Helen Bernstein Book Award for Excellence in Journalism, and is a four-time nominee for the National Book Critics Circle Award. In 2002 he won the National Magazine Award for Excellence in Reporting for "The Crash of EgyptAir 990," which appeared in the *Atlantic Monthly.* Two years later, his groundbreaking reportage on the Columbia Space Shuttle disaster, "Columbia's Last Flight," was awarded the same prize. Langewiesche lives in France and California.

Jonah Lehrer, twenty-five, is editor at large for *Seed* magazine. A graduate of Columbia University and a Rhodes scholar, Lehrer has also written for *Nature, New Scientist,* and NPR. His first book, *Proust Was a Neuroscientist,* will be published by Houghton Mifflin in 2007.

Michael D. Lemonick left the staff of *Time* magazine in February 2007 to become a *Time* contributor and freelance writer. During his nearly twenty-one years at *Time,* he produced fifty cover stories on science, medicine, and the environment, along with innumerable smaller stories. He has also written on a freelance basis for *Discover, Audubon, People, Playboy,* and the *Washington Post.* Lemonick is the author of three books: *The Light at the Edge of the Universe* (1993), *Other Worlds* (1997), and *Echo of the Big Bang* (2003). His fourth book, *The Seventh Planet: William Herschel and the Discovery of Uranus* will be published in 2008 by James Atlas Books as part of its Great Discoveries Series. In addition to writing, Lemonick teaches at Princeton and at New York University.

Jeffrey A. Lockwood came to the writing life through the back door. He earned a B.S. in biology from New Mexico Tech and a Ph.D. in entomology from Louisiana State University. Originally hired as an assistant professor of entomology at the University of Wyoming, he metamorphosed into a professor of natural sciences and humanities and transferred to the Department of Philosophy and the MFA program in creative writing. He teaches workshops in nature and spiritual/religious writing, along with courses in environmental and natural resource ethics and the philosophy of ecology. His writings have been honored with a Pushcart Prize and a John Burroughs Award. His current book project is *Six-Legged Soldiers: The Use of Insects as Weapons of War and Terror.*

Lynn Margulis, Distinguished University Professor in the Department of Geosciences at the University of Massachusetts in Amherst, is dedicated to laboratory and scholarly work and fieldwork (in Cape Cod, Spain, and Mexico). As a teacher at many levels, she identifies herself as a nature-loving evolutionist. She earned her A.B. in liberal arts from the University of Chicago, where she was not permitted to major in anything — a fact to which she attributes her excellent education and love of science. She received an M.S. in zoology and genetics from the splendid biology program at the University of Wisconsin, Madison, and her Ph.D. in genetics from the University of California at Berkeley. She was honored with a National Medal of Science by President Clinton in 2000 and notes that she would have refused to accept it from George Bush.

Steve Olson is a writer in Washington, D.C., and the author most recently of *Count Down: Six Kids Vie for Glory at the World's Toughest Math Competition.* His 2002 book, *Mapping Human History: Discovering the Past Through Our Genes,* was nominated for the National Book Award. He has written for the *Atlantic Monthly, Wired, Science, Scientific American, Slate,* and other magazines. In 2004 he and two coauthors published an article in *Nature* showing that the most recent common genealogical ancestor of everyone on Earth today probably lived between two and three thousand years ago.

Stephen W. Pacala is a biologist and the director of the Princeton Environmental Institute. He also directs the Carbon Modeling Consortium, which works to understand all aspects of the global carbon cycle. The group is composed of ecologists, physical and biological oceanographers, and atmospheric scientists and investigates issues ranging from the effects of global vegetation on climate to the large-scale measurement of natural and anthropogenic greenhouse gas emissions.

Michael Perry is the author of the memoirs *Population 485: Meeting Your Neighbors One Siren at a Time* and *Truck: A Love Story* and of the essay collection *Off Main Street.* A volunteer firefighter and emergency medical responder in rural Wisconsin, he lives online at www.sneezingcow.com.

Heather Pringle has spent twenty-five years writing about archaeology and anthropology. During her lengthy career, she has written for magazines as diverse as *Science, Stern, Geo, New Scientist, Discover, National Geographic Traveler, Islands, Readers Digest,* and *Canadian Geographic.* Her articles have won numerous awards, including the prestigious Science Journalism Award from the American Association for the Advancement of Science. She currently lives in Vancouver, Canada.

Jonathan Rauch is a senior writer and columnist for *National Journal* magazine, a correspondent for the *Atlantic Monthly,* and a guest scholar at the

Brookings Institution. He is the author of several books and many articles on public policy, culture, and economics. His latest book is *Gay Marriage: Why It Is Good for Gays, Good for Straights, and Good for America* (2004). It makes the case that, properly implemented, same-sex marriage would benefit not only gay couples but society and, equally important, the institution of marriage itself. Although much of Rauch's writing has been on public policy, he has also written on topics as widely varied as adultery, agriculture, economics, height discrimination, biological rhythms, and animal rights. He received the 2005 National Magazine Award for columns and commentary.

Michael Rosenwald is a staff writer for the *Washington Post.* A former finalist for the National Magazine Award in feature writing, Rosenwald has also written for *The New Yorker, Esquire, Popular Science, Smithsonian, Discover, Creative Nonfiction,* and *Tin House.* His writing was recently featured in the anthology *The Best Creative Nonfiction, Vol. 1.* Rosenwald was previously a reporter for the *Boston Globe.* He studied narrative nonfiction in the MFA program at the University of Pittsburgh. He lives with his wife in Germantown, Maryland.

Bonnie J. Rough has won the McKnight Artist Fellowship for Writers, the Iowa Review Award, and the Annie Dillard Award for Creative Nonfiction. Her essays have appeared recently in *Modern Love: 50 True and Extraordinary Tales of Desire, Deceit, and Devotion* (2007), the *New York Times, Brevity,* and *Isotope.* She lives in Minneapolis, where she teaches at the Loft Literary Center and is at work on two books: a collection of essays on flight and a memoir about genetic heredity.

Robert M. Sapolsky is a professor of biology and neurology at Stanford University and a research associate with the Institute of Primate Research, National Museum of Kenya. A neuroendocrinologist, he focuses his research on the effects of stress. Two of his books, *Why Zebras Don't Get Ulcers: An Updated Guide to Stress, Stress-Related Disease, and Coping* and *The Trouble with Testosterone: And Other Essays on the Biology of the Human Predicament* were Los Angeles Times Book Club finalists. He has received numerous honors and awards for his work, including a MacArthur Fellowship and an Alfred P. Sloan Fellowship. He is a regular contributor to *Discover* and *Science.*

John Seabrook has been a staff writer at *The New Yorker* since 1993. He is the author of *Deeper: My Two-Year Odyssey in Cyberspace* (1997) and *Nobrow: The Culture of Marketing the Marketing of Culture* (2000). His work has also appeared in *Harper's Magazine,* the *Nation, Vanity Fair, Vogue, Travel + Leisure,* *and* the *Village Voice.* He has taught narrative nonfiction writing at Princeton University and lives in New York City.

Bill Sherwonit, who was born in Bridgeport, Connecticut, has called Alaska home since 1982. He has contributed essays and articles to a wide variety of newspapers, magazines, journals, and anthologies and is the author of eleven books about Alaska, including *Living with Wildness: An Alaskan Odyssey,* which will be published this winter by the University of Alaska Press. He also coedited *Travelers' Tales Alaska.* Sherwonit lives in Anchorage's Turnagain area, where he writes about the wildness to be found in the state's urban centers as well as its far reaches. His Web site is www.billsherwonit.alaskawriters.com.

Michael Shnayerson has been a contributing editor of *Vanity Fair* for many years and has written nearly one hundred stories for the magazine on subjects ranging from politics to pop stars. For fun, he has also written books on subjects outside *Vanity Fair*'s realm: on electric cars (*The Car That Could: The Story of GM's Revolutionary Electric Vehicle*) and antibiotic resistance (*The Killers Within: The Deadly Rise of Drug-Resistant Bacteria*). "The Rape of Appalachia" is a sort of distilled, early version of his next book, *Coal River: An American Story,* to be published in January 2008. Shnayerson lives in Bridgehampton, New York.

Meredith F. Small is a writer and professor of anthropology at Cornell University. She has been writing on science and nature for over twenty years, publishing features, essays, and news articles in all the major magazines and newspapers. She is best known for her book *Our Babies, Ourselves; How Biology and Culture Shape the Way We Parent,* which has become a cult classic among new parents. Her latest book is *The Culture of Our Discontent: Beyond the Medical Model of Mental Illness,* which looks at how other cultures define and treat mental illness. She is currently working on a book about the science, nature, and history of wool, everything from sheep to sweater. Sheep shearing, she has recently learned, is not for the squeamish.

Robert H. Socolow, professor of mechanical and aerospace engineering at Princeton University, teaches in both the School of Engineering and Applied Science and the Woodrow Wilson School of Public and International Affairs. With the ecologist Stephen Pacala, Socolow leads the university's Carbon Mitigation Initiative. His research focuses on technology and policy for fossil fuels under climate constraints. He was awarded the 2003 Leo Szilard Lectureship Award by the American Physical Society "for leadership in establishing energy and environmental problems as legitimate research fields for physicists, and for demonstrating that these broadly defined problems can be addressed with the highest scientific standards." Socolow earned a B.A. in 1959 (summa cum laude) and a Ph.D. in theoretical high energy physics in 1964 from Harvard University.

Neil deGrasse Tyson was born and raised in New York City, where he was educated in the public schools clear through his graduation from the Bronx High School of Science. Tyson went on to earn his B.A. in physics from Harvard and his Ph.D. in astrophysics from Columbia. His professional research interests are broad and include star formation, exploding stars, dwarf galaxies, and the structure of the Milky Way. Tyson obtains his data from the Hubble Space Telescope, as well as from telescopes in California, New Mexico, Arizona, and the Andes Mountains of Chile. In addition to publishing in dozens of professional publications, he has written and continues to write for the public. He is a monthly essayist for *Natural History,* writing the column titled "Universe." His latest book, *Death by Black Hole: And Other Cosmic Quandaries,* is a *New York Times* best-selling collection of his favorite essays from the past eleven years.

Ethan Watters is a freelance writer who lives in San Francisco with his wife and two children. His youngest, Toby, was in utero while he was researching the story on the effects of epigenetics on early fetal development.

Other Notable Science and Nature Writing of 2006

SELECTED BY TIM FOLGER

JERRY ADLER
Finding a Home in the Cosmos. *Smithsonian,* July.
PHILIP ALCABE
The Ordinariness of AIDS. *The American Scholar,* Summer.
IVAN AMATO
Bar Coding Life. *Chemical & Engineering News,* January 16.
ANDO ARIKE
Owning the Weather. *Harper's Magazine,* January.
PHILIP ARMOUR
Bloody Business. *Outside,* May.

JOHN PERRY BARLOW
Is Cyberspace Still Anti-Sovereign? *California,* March–April.
RICK BASS
Return of the Black Rhino. *On Earth,* Spring.
JEREMY BERNSTEIN
The Secrets of the Bomb. *The New York Review of Books,* May 25.
BURKHARD BILGER
The Search for Sweet. *The New Yorker,* May 22.
GEORGE BLACK
Patagonia Under Siege. *On Earth,* Fall.
KENNETH BROWER
Disturbing Yosemite. *California,* May–June.
ALAN BURDICK
On Excellence. *Ascent,* Fall.

INGFEI CHEN
Born to Run. *Discover,* May.

GAY DALY
Hundreds of Man-made Chemicals . . . *On Earth,* Winter.
BILL DOUTHITT
Beautiful Stranger. *National Geographic,* December.
MADELINE DREXLER
The People's Epidemiologists. *Harvard Magazine,* March–April.

BRIAN FAGAN
Archaeology: The Next 50 Years. *Archaeology,* September–October.
VICTORIA FINLAY
The Tears of Trees. *Orion,* July–August.

NANCY GIBBS
Stem Cells: The Hope and the Hype. *Time,* August 7.
DANIEL GLICK
Leader of the Pack. *Audubon,* March–April.
FRED GUTERL
Journey to the Outer Limits. *Discover,* March.

EDWARD HOAGLAND
Miles from Nowhere. *The American Scholar,* Summer.
JOSEPH HOOPER
Mr. Hard Cell. *Popular Science,* August.
ALEX HUTCHINSON
The Next Atomic Age. *Popular Mechanics,* October.

ROBERT IRION
The Planet Hunters. *Smithsonian,* October.

GEORGE JOHNSON
Worlds Apart. *Tricycle,* Spring.

JENNIFER KAHN
Itchy. *Outside,* September.
JEFFREY KLUGER
Why We Worry About the Things We Shouldn't . . . and Ignore the Things We Should. *Time,* December 4.

ANDREW LAWLER
Damming Sudan. *Archaeology,* November–December.
Inside the Restoration of the Amiriya Madrassa. *Discover,* April.
ANTONIO LAZCANO
The Origins of Life. *Natural History,* February.
SHARON LEVY
Mammoth. *On Earth,* Winter.
The Vanishing. *On Earth,* Summer.

BRUNO MADDOX
Blinded by Science. *Discover,* December.
ANDREW MARSHALL
Making a Killing. *Bulletin of the Atomic Scientists,* March–April.
JOHN G. MITCHELL
When Mountains Move. *National Geographic,* March.

ANNALEE NEWITZ
Your Secret Life Is Ready. *Popular Science,* September.
MICHELLE NIJHUIS
Wrecking the Rockies. *On Earth,* Summer.

COREY POWELL
My Three Einsteins. *Discover,* October.

STEPHEN CHRISTOPHER QUINN
The Worlds Behind the Glass. *Natural History,* April.

PAUL RAFFAELE
Sleeping with Cannibals. *Smithsonian,* September.

NICK SCHWELLENBACH AND PETER D. H. STOCKTON
Nuclear Lock. *Bulletin of the Atomic Scientists,* November–December.
JAMES SHREEVE
The Greatest Journey. *National Geographic,* March.
CHARLES SIEBERT
Unintelligent Design. *Discover,* March.

PAUL THEROUX
Living with Geese. *Smithsonian,* December.
JAY TOLSON
Is There Room for the Soul? *U.S. News & World Report,* October 23.
MIKE TONER
Impossibly Old America? *Archaeology,* May–June.
WELLS TOWER
The Thing with Feathers. *Outside,* March.

CAREL VAN SCHAIK
Why Are Some Animals So Smart? *Scientific American,* April.
ALEX VILENKIN
Beyond the Big Bang. *Natural History,* July–August.

TODD WILKINSON
Winterkeeper. *Audubon,* January–February.

TOM YULSMAN
Snow Daze. *Audubon,* January–February.

THE B·E·S·T AMERICAN SERIES®

THE BEST AMERICAN SHORT STORIES® 2007. STEPHEN KING, editor, HEIDI PITLOR, series editor. This year's most beloved short fiction anthology is edited by Stephen King, author of sixty books, including *Misery, The Green Mile, Cell,* and *Lisey's Story,* as well as about four hundred short stories, including "The Man in the Black Suit," which won the O. Henry Prize in 1996. The collection features stories by Richard Russo, Alice Munro, William Gay, T. C. Boyle, Ann Beattie, and others.

ISBN-13: 978-0-618-71347-9 • ISBN-10: 0-618-71347-6 $28.00 CL
ISBN-13: 978-0-618-71348-6 • ISBN-10: 0-618-71348-4 $14.00 PA

THE BEST AMERICAN NONREQUIRED READING™ 2007. DAVE EGGERS, editor, introduction by SUFJAN STEVENS. This collection boasts the best in fiction, nonfiction, alternative comics, screenplays, blogs, and "anything else that defies categorization" (*USA Today*). With an introduction by singer-songwriter Sufjan Stevens, this volume features writing from Alison Bechdel, Scott Carrier, Miranda July, Lee Klein, Matthew Klam, and others.

ISBN-13: 978-0-618-90276-7 • ISBN-10: 0-618-90276-7 $28.00 CL
ISBN-13: 978-0-618-90281-1 • ISBN-10: 0-618-90281-3 $14.00 PA

THE BEST AMERICAN COMICS™ 2007. CHRIS WARE, editor, ANNE ELIZABETH MOORE, series editor. The newest addition to the Best American series—"A genuine salute to comics" (*Houston Chronicle*)—returns with a set of both established and up-and-coming contributors. Edited by Chris Ware, author of *Jimmy Corrigan: The Smartest Kid on Earth,* this volume features pieces by Lynda Barry, R. and Aline Crumb, David Heatley, Gilbert Hernandez, Adrian Tomine, Lauren Weinstein, and others.

ISBN-13: 978-0-618-71876-4 • ISBN-10: 0-618-71876-1 $22.00 CL

THE BEST AMERICAN ESSAYS® 2007. DAVID FOSTER WALLACE, editor, ROBERT ATWAN, series editor. Since 1986, *The Best American Essays* has gathered outstanding nonfiction writing, establishing itself as the premier anthology of its kind. Edited by the acclaimed writer David Foster Wallace, this year's collection brings together "witty, diverse" (*San Antonio Express-News*) essays from such contributors as Jo Ann Beard, Malcolm Gladwell, Louis Menand, and Molly Peacock.

ISBN-13: 978-0-618-70926-7 • ISBN-10: 0-618-70926-6 $28.00 CL
ISBN-13: 978-0-618-70927-4 • ISBN-10: 0-618-70927-4 $14.00 PA

THE BEST AMERICAN MYSTERY STORIES™ 2007. CARL HIAASEN, editor, OTTO PENZLER, series editor. This perennially popular anthology is sure to appeal to mystery fans of every variety. The 2007 volume, edited by best-selling novelist Carl Hiaasen, features both mystery veterans and new talents. Contributors include Lawrence Block, James Lee Burke, Louise Erdrich, David Means, and John Sandford.

ISBN-13: 978-0-618-81263-9 • ISBN-10: 0-618-81263-6 $28.00 CL
ISBN-13: 978-0-618-81265-3 • ISBN-10: 0-618-81265-2 $14.00 PA

THE BEST AMERICAN SERIES®

THE BEST AMERICAN SPORTS WRITING™ 2007. DAVID MARANISS, editor, GLENN STOUT, series editor. "An ongoing centerpiece for all sports collections" (*Booklist*), this series stands in high regard for its extraordinary sports writing and topnotch editors. This year David Maraniss, author of the critically acclaimed biography *Clemente*, brings together pieces by, among others, Michael Lewis, Ian Frazier, Bill Buford, Daniel Coyle, and Mimi Swartz.

ISBN-13: 978-0-618-75115-0 • ISBN-10: 0-618-75115-7 $28.00 CL
ISBN-13: 978-0-618-75116-7 • ISBN-10: 0-618-75116-5 $14.00 PA

THE BEST AMERICAN TRAVEL WRITING™ 2007. SUSAN ORLEAN, editor, JASON WILSON, series editor. Edited by Susan Orlean, staff writer for *The New Yorker* and author of *The Orchid Thief*, this year's collection, like its predecessors, is "a perfect mix of exotic locale and elegant prose" (*Publishers Weekly*) and includes pieces by Elizabeth Gilbert, Ann Patchett, David Halberstam, Peter Hessler, and others.

ISBN-13: 978-0-618-58217-4 • ISBN-10: 0-618-58217-7 $28.00 CL
ISBN-13: 978-0-618-58218-1 • ISBN-10: 0-618-58218-5 $14.00 PA

THE BEST AMERICAN SCIENCE AND NATURE WRITING™ 2007. RICHARD PRESTON, editor, TIM FOLGER, series editor. This year's collection of the finest science and nature writing is edited by Richard Preston, a leading science writer and author of *The Hot Zone* and *The Wild Trees*. The 2007 edition features a mix of new voices and prize-winning writers, including James Gleick, Neil deGrasse Tyson, John Horgan, William Langewiesche, Heather Pringle, and others.

ISBN-13: 978-0-618-72224-2 • ISBN-10: 0-618-72224-6 $28.00 CL
ISBN-13: 978-0-618-72231-0 • ISBN-10: 0-618-72231-9 $14.00 PA

THE BEST AMERICAN SPIRITUAL WRITING™ 2007. PHILIP ZALESKI, editor, introduction by HARVEY COX. Featuring an introduction by Harvey Cox, author of the groundbreaking *Secular City*, this year's edition of this "excellent annual" (*America*) contains selections that gracefully probe the role of faith in modern life. Contributors include Robert Bly, Adam Gopnik, George Packer, Marilynne Robinson, John Updike, and others.

ISBN-13: 978-0-618-83333-7 • ISBN-10: 0-618-83333-1 $28.00 CL
ISBN-13: 978-0-618-83346-7 • ISBN-10: 0-618-83346-3 $14.00 PA